Project AIR FORCE

T0167447

THE VIRTUAL COMBAT AIR STAFF

THE PROMISE OF INFORMATION TECHNOLOGIES

Arthur F. Huber
Philip S. Sauer
J. Lawrence Hollett
Kenneth Keskel
William L. Shelton, Jr.
John T. Dillaplain

Prepared for the United States Air Force

RAND

The purpose of this study was to investigate the nature of the future combat air staff in the context of air war in the information age. The thrust of the research centered on assessing the issues inherent in employing the concept of "virtuality" to make the deployed combat air staff's operations more efficient and effective.

Within the confines of this study, *virtuality* refers to the concept that not all elements of a staff may be physically located in the same place, that communications technology may allow for the retrieval of information resources from diverse centers of responsibility, and that staff assets may be re-absorbed into host centers after the cessation of hostilities. Application of this concept to the military realm has been suggested, but not rigorously investigated. Further, the boundaries of the debate about the applicability of virtuality to military staff operation have not been systematically framed. In this context, the authors chose to limit the application of virtuality to a combat air staff; that is, that staff needed to plan, coordinate, and execute an air campaign plan in support of theater operations. It is the hope of the authors that this research may energize discussion on this issue and contribute towards establishing the appropriate context for further study.

Research at RAND conducted under the Project AIR FORCE program is normally proposed, approved, and conducted through a formal review process that involves obtaining guidance from and interaction with an Air Force customer. This study was conducted as a purely independent endeavor by Air Force Fellows resident at RAND as part of a tour of duty to fulfill in-residence Intermediate Service School requirements. Although materiel and intellectual support was provided by the RAND staff, the opinions expressed herein belong strictly to the authors and do not represent those of RAND or the U.S. Air Force.

This report is provided for your reading and deliberation. This report is not about answers. Rather, it wrestles with many of the questions now facing the Air Force as it struggles to adapt to the rapid changes imposed by the march of technology. If you desire to comment on this report, please send your comments directly to Project AIR FORCE, RAND, 1700 Main Street, Santa Monica, California, 90407-2138.

PROJECT AIR FORCE

Project AIR FORCE, a division of RAND, is the Air Force federally funded research and development center (FFRDC) for studies and analyses. It provides the Air Force with independent analyses of policy alternatives affecting the development, employment, combat readiness, and support of current and future aerospace forces. Research is being performed in three programs: Strategy and Doctrine; Force Modernization and Employment; and Resource Management and System Acquisition. Project AIR FORCE is operated under contract F49642-96-C-0001 between the Air Force and RAND.

In 1996, Project AIR FORCE is celebrating 50 years of service to the United States Air Force. Project AIR FORCE began in March 1946 as Project RAND at Douglas Aircraft Company, under contract to the Army Air Forces. Two years later, the project became the foundation of a new, private nonprofit institution to improve public policy through research and analysis for the public welfare and security of the United States—what is known today as RAND.

CONTENTS

FIGURES

SUMMARY

The principal objective of this study was to explore whether the possibilities provided by advancing technology and alternative organizational models, as they might affect the deployed combat air staff in future military campaigns, would permit a distributed network organization of effort, operating in the virtual realm. The thrust of the research centered on assessing the issues inherent in employing the concept of "virtuality" to make deployed combat air staff operations more efficient and effective. A combat air staff is that staff required to perform the necessary actions to plan, coordinate, and execute an air campaign plan in support of military objectives. In essence, we postulate the need for a hybrid organizational structure that takes advantage of the best parts of the traditional military hierarchical and network information models. Within the confines of this study, *virtuality* refers to the concept that not all elements of a staff may be physically located in the same place, that communications technology may allow for the retrieval of information resources from diverse centers of responsibility, and that staff assets may be reabsorbed into host centers after the cessation of hostilities.

A set of core questions served to shape this inquiry: How did the current standards for military staffs evolve? How will projected technological capabilities and advancements provide opportunities for creating "virtual" staff organizations to better manage and execute future combat operations? What can we learn about "virtual" organizations from the commercial sector? What questions should be asked to properly frame the derivation of future combat air staff requirements?

The results of this research indicate that the rapid advances now progressing within the technological realm, as well as within organizational theory and practice, bode a different paradigm for the future combat air staff. Whereas the past has seen a general trend of growth in staff size and dependence on hierarchical constructs, the future may actually see something approaching the opposite. Future deployed staffs may be smaller and geared toward working within a network architecture that places less emphasis on hierarchical relationships. This evolution may be motivated as much by a shift to network operations by potential adversaries as through a natural application of the capabilities inherent in emerging technologies.

The key to understanding the foundation of the ideas in this report is to recognize the continuing need of the battlefield commander to maximize his "presence," or awareness of events, across the entirety of the military theater. Further, the desire of military commanders to reduce the numbers of people deployed into a combat region and to avoid placing them in harm's way militates for application of technologies and capabilities that permit execution of the air campaign with a reduced number of people. Recognition of this intrinsic—and increasingly important—military function allows one to perceive that a confluence of requirements, technological advancements, and virtual organizational constructs may provide the impetus for the creation of a new standard for implementing the staff function.

A review of the historical evolution of the military staff provides key insights into how to view change in this realm. It appears that, despite a glacial rate of improvement during antiquity followed by little change (if not outright regression) during the Middle Ages, the modern era has been one of increasingly rapid change in military staff organization and functional employment. Nothing in the record indicates that the rate of change will slacken, although it is entirely possible that the nature of these changes may itself evolve. Technology has been and will continue to be a driver for this evolution. Nevertheless, the need to conduct ever more complex combat operations will continue to influence adaptation within staff organizations strongly.

The elements of the technological revolution that enable consideration of virtual concepts for creating the military staff of the future are many and varied. However, fundamental to this revolution have been the phenomenal advances made in the capabilities of the microchip. By leaps and bounds, the microchip has advanced in every feature of its performance that matters, from physical size to processing speed to memory capacity. Along with this progress have come improvement in data rates for communications systems, development of expert systems, artificial intelligence, and neural nets, as well as sophisticated displays and speech recognition systems. The mobility of computer and communications hardware has greatly improved, and security issues are being tackled. Our conclusion is that the necessary technologies to make the virtual staff construct a reality are either in work today or can be reasonably expected to come to fulfillment if historical trends continue.

The revolution in information technology and telecommunications combined with intense global competition has fostered immense changes in the commercial sector. Large companies have restructured, and small ones have sprouted that exercise greater influence than their size would portend. Within this environment, a number of companies—both large and small—have had success implementing the principles of the virtual organization. These principles have manifested themselves in a variety of ways, such as corporations built on ad hoc or informal relationships, mobile workplaces, global production networks, and telecommuting. Companies that use these principles are demonstrating greater "agility" resulting in significant bottom-line success. This "agility" allows a company to better leverage core competencies to take advantage of niche or fleeting opportunities within the marketplace and thus derive a competitive advantage. "Reengineering" initiatives have also resulted in "flatter" organizational hierarchies, employment of cross-functional teams, and an emphasis on streamlining work, retaining only value-added processes. The challenge for the U.S. military is to develop new organizational structures that achieve the efficiencies and creativity businesses have gained in the virtual and reengineered environments, while at the same time retaining the elements of the traditional, hierarchical, command and control system (e.g., discipline, morale, tradition) essential for operations in the combat arena. Such a hybrid organizational structure is at the foundation of this study and may be the topic requiring debate within the military as a whole to chart the course into the information age.

As business and social institutions increasingly adopt virtual princi-ples, it is most opportune for those concerned with shaping and leading our military forces to ask how such developments will affect the future of the combat air staff. They will need to deal with such issues as how the concept of command should change, what organi-zational construct will best facilitate these types of operations, how operational campaign planning and force management can be made more efficient, how training should be conducted to ensure opti-mized employment of virtual concepts, to what extent information security shortcomings can be overcome, and how virtuality might change relationships within joint forces, among coalition partners, and with the media. These are but some of the challenges facing our leadership, and it does not appear that circumstances will tolerate much delay in coming to grips with them. Only by focusing on core staff functions and applying a disciplined analysis to their proper placement within the emerging virtual paradigm can such issues be properly addressed.

ACKNOWLEDGMENTS

The authors express their appreciation to the many individuals who made this research effort possible. From technical editing and report typing to research support and intellectual guidance, many individual contributions were made to ensure that the fruits of this endeavor were published and disseminated. While we cannot thank all contributors by name, we want everyone who assisted to realize that they were instrumental in making this effort possible. For this cheerful and professional support, we will always be grateful.

Several individuals on the RAND staff deserve special recognition for their contributions to this report. Myron Hura served as mentor, helping guide the research and documentation of this project. Without his inspiration, it is doubtful that we would have had the fortitude to see this project through to its conclusion. Glenn Buchan, Douglas Merrill, and Gary McLeod also provided extensive review and comments, greatly improving the report's content and structure. The contributions of these individuals were key to our coming to grips with the many facets of the virtual approach to communicating among people and organizing them to accomplish large-scale tasks. Finally, Phyllis Gilmore, Emily Rogers, and Virginia Tura provided invaluable assistance "above and beyond" their normal duties helping to give the final product a quality appearance.

Although numerous people contributed their time and effort in support of this report, we remain responsible for the judgments and observations contained herein.

ABBREVIATIONS

2-D	Two-dimensional
3-D	Three-dimensional
4-D	Four-dimensional
AAA	Anti-aircraft artillery
ACC	Air Combat Command
AFB	Air force base
AI	Artificial intelligence
ATM	Asynchronous transfer mode
ATO	Air Tasking Order
AWACS	Airborne Warning and Control System
BDA	Battle damage assessment
BISDN	Broadband Integrated Services Digital Network
C^2	Command and control
C^3	Command, control, and communications
C^3I	C^3 and intelligence
C^4	Command, control, communications and computers
C^4I	C^4 and intelligence
C^4IFTW	C^4I for the Warrior
CAFMS	Computer Aided Force Management System
CAS	Close Air Support
CD	Compact disk
CENTCOM	Central Command
CENTAF	Central Command Air Forces
CERT	Computer Emergency Response Team
CIA	Central Intelligence Agency
CINC	Commander in chief
CNN	Cable News Network

CONUS	Continental United States
COTS	Commercial off the shelf
CTULD	Countdown time until landing display
DARPA	Defense Advanced Research Projects Agency
DBS	Direct broadcast service
DFC	Distinguished Flying Cross
DG	Distinguished graduate
DIA	Defense Intelligence Agency
DISA	Defense Information Systems Agency
DoD	Department of Defense
DSCS III	Defense Satellite Communications System III
DSR	Daily situation report
ECM	Electronic countermeasures
EHF	Extremely high frequency
EINet	Enterprise Integration Network
e-mail	Electronic mail
Gbps	Gigabits per second
GBS	Global Broadcast Services
GHz	Gigahertz
HDTV	High-definition television
HUD	Head-up display
HUMINT	Human intelligence
IC	Integrated circuit
IFE	In-flight emergency
INMARSAT	International Maritime Satellite
IP	Instructor pilot
JC2WC	Joint Command and Control Warfare Center
JCS	Joint Chiefs of Staff
JFACC	Joint Forces Air Component Commander
JFC	Joint Force Commander
JSAS	JFACC Situational Awareness System
JSTARS	Joint Surveillance [and] Target Attack Radar System
JTF	Joint Task Force
JTIDS	Joint Tactical Information Data System
Kbps	Kilobits per second
LEO	Low earth orbit
MAP	Master attack plan
Mbps	Megabits per second
MHz	Megahertz

MILSTAR	Military Strategic and Tactical Relay (communications satellites)
MIPS	Millions of instructions per second
MIT	Massachusetts Institute of Technology
NAPIL	North Africa Pan-Islamic League
NATO	North Atlantic Treaty Organization
NSA	National Security Agency
OPSEC	Operations security
PACAF	Pacific Air Forces
PC	Personal computer
PGM	Precision guided munition
R&D	Research and development
R&R	Rest and relaxation
ROM	Read-only memory
SAM	Surface-to-air missile
SAR	Synthetic aperture radar
SHF	Super high frequency
SITREP	Situation report
SONET	Synchronous Optical Network
TDY	Temporary duty
tps	Transactions per second
UAV	Unmanned aerial vehicle
UFO	UHF Follow-On
UHF	Ultra high frequency
USAF	United States Air Force
USAFE	United States Air Forces Europe
VCJCS	Vice Chairman Joint Chiefs of Staff
VCM	Visual countermeasures
VHS	Video Handling System (Videotape Format)
VTC	Video teleconferencing
WNN	World News Network

Chapter One

INTRODUCTION

Changes brought forth by the information age continue to modify our projections of how combat air operations will be conducted in the future. The quantification of those changes is a large, worthwhile objective but is beyond the scope of this study. Instead, we focused qualitatively on the use of a "virtual" combat air staff to support a more flexible, leaner Air Force of the future, taking advantage of anticipated computing and communications advances that may require adapting to the accompanying network organizational construct. Our attention centered on the idea of describing critical issues of a virtual combat air staff. This staff is virtual because not all of the elements of the staff may be physically located in the same place, communications technology allows for the retrieval of information resources from diverse centers of responsibility, and staff assets may be reabsorbed into host organizations after the cessation of hostilities. As such, this report seeks to provide a framework within which a discussion of the concept of a virtual combat air staff, as one part of a hybrid organizational structure, can begin.

HIERARCHICAL VERSUS NETWORK ORGANIZATIONAL MODELS

A hierarchical organizational model is at the heart of every nation-state's military. This organizational form is characterized by a vertical flow of decisionmaking and information, with the focus arranged at the top of the flow. The hierarchical military organization has proven to be an adequate construct for waging industrial-age warfare against a similarly organized opponent, even though there are inherent limitations to this organization. Two of those limitations include poor information transfer mechanisms and the inflexibility to take the maximum advantage of newer technologies. The U.S. military is no exception in this regard.

However, new organizational structures are emerging to allow humans to better interact and communicate in the information age. One clear indication is that networks are becoming preferred organizational structures, sometimes at the expense of hierarchical constructs. The network form of organization contrasts directly with the hierarchical form. A network is characterized by multiple nodes interacting directly with each other, fully exchanging information without regard for lines of command or organizational structure. While this form permits rapid information transfer, the network form has poor or ill-defined control mechanisms, strengths that the hierarchical form of organization embodies. This poses a clear challenge to the military, because military organizations within nation-states are typically built on hierarchical organizational models. Such an organization has been used because of its vertical flow of control, facilitating dissemination of orders from top to bottom and ensuring compliance from bottom to top in a rapid, efficient manner. In today's military, there is a need for both clear command and control (C^2) and rapid information exchange, driving the need for both hierarchical and network forms.

Operations other than war and low-intensity conflict appear to be increasing in frequency. Adversaries in these types of operations pose different problems:

> Most adversaries that the United States and its allies face in the realms of low-intensity conflict—international terrorists, guerrilla insurgents, drug cartels, ethnic factions, as well as racial gangs, and smugglers of illegal aliens—are all organized like networks (although their leadership may be quite hierarchical). Perhaps a reason that military and police institutions keep having difficulty engaging in low-intensity conflicts is because they are not meant to be fought by institutions. The lesson: Institutions can be defeated by networks. It may take networks to counter networks. The future may belong to whoever masters the network form. (Arquilla and Ronfeldt, 1992, p. 17.)

This report examines the network form of interaction and the promise for application of virtual principles, possibly within a mixed hierarchical C^2 construct and a network execution form. This new paradigm may be what is necessary to be able to respond to forms of conflict based on a distributed and agile target set, as well as on asymmetrical strategies.

EXPEDITIONARY WARFARE: "DEPLOYING" A COMBAT AIR STAFF

A number of factors point to an "expeditionary" U.S. Air Force element as a critical component in the conduct of future air war. Significant reductions in permanent forward basing, deployments outside of the traditional European and Pacific locations, and involvement in military operations other than war in the Third World are the most commonly documented aspects of a "global reach, global power" strategy.[1]

The nature of the regional Unified Commands includes responsibilities for such broad and diverse areas as to limit specialization by assigned combat air staff. In the past, specialists physically augmented the deploying staff to fill gaps in expertise. Technological advances may enable these gaps to be filled virtually, using advanced communications and information processing to bring expertise and data into a theater without physically deploying additional personnel. As a result, use of a virtual combat air staff concept can provide the flexibility to minimize the resources physically deployed, permit addition of staff expertise as needed through communications connectivity, allow the participation of the most competent staff members, and achieve efficiencies by virtue of allowing a larger percentage of personnel to remain at their home bases. Presumably, the in-theater air combat commander would still be able to retain direct C^2 while using this "reachback" concept to access the various resources and information required to execute the air campaign plan.

---

[1]For a good summary of this strategy, see Lynch (1995), pp. 24–31.

DEPLOYING TO WAGE WAR

Setting: U.S. Central Command (CENTCOM) Headquarters, MacDill Air Force Base, Florida, August 1990.

The CENTCOM staff faced a daunting task during August 1990: successful coordination of the largest deployment of U.S. military forces since the Korean War. A number of pressing immediate requirements influenced their efforts to perform this task: the overarching need to maintain surveillance of a massive Iraqi military force in the midst of invading and securing Kuwait (and possibly extending the invasion into the oil-rich region of Saudi Arabia); an equally critical necessity to closely monitor and control a dangerous blockade by U.S. naval forces in the Gulf; and, as part of CENTCOM's wider geographic responsibility, the tracking of a developing hurricane off the eastern coast of Africa. A critical component of this force deployment question directly affected the members of the CENTCOM staff—who, among headquarters personnel, would deploy, and when?

The CENTCOM current operations center, filled with powerful information systems, could not easily be replicated in the Saudi Ministry of Defense in Riyadh. Each CENTCOM staff member's computer displayed a dazzling array of information ranging from the status of transport ships headed for Saudi ports to the current order of battle information for various Iraqi military forces. The Cable News Network (CNN) projected from large overhead color monitors. An electronic map on the center wall displayed tracks of the U.S. carrier battle groups in the region and the transit of merchant vessels. The staff could communicate directly with forward elements in Riyadh and command ships in international waters throughout the region. Situation charts, constantly updated on computer workstations, could be forwarded via fax to the Joint Staff in Washington and other commands. Advanced communications capability provided both direct contact with, and a "virtual presence" in, information centers and command centers across the globe.[2]

---

[2]All descriptions of the physical layout, capabilities and deployment issues are taken directly from Macedonia (1992), pp. 34–35.

The command structure of CENTCOM and its supporting information systems shaped the deployment debate. CENTCOM, a regional unified command with military warfighting responsibilities from Egypt to Pakistan and from the northern borders of Iraq and Iran down the eastern coast of Africa, did not physically "own" any combat military forces. Its component forces resided in other major commands during peacetime. But with CENTCOM's structure molded on the traditional military hierarchical C^2 model, the question was how to manage the massive information required to organize the deployment and employment of modern military forces. As Major Michael Macedonia relates,

> The irony of the modern age was that the CENTCOM commander, at the outset of the crisis, could gather more intelligence information and more effectively control his forces 7,000 miles away from the theater of operations than from the Saudi Ministry of Defense. The command could deploy too early, only to have to call MacDill to find out what was going on in the theater. Not only did CENTCOM have to await the build up of its forces but also the development of a communications infrastructure to fight the war. (Macedonia, 1992, p. 35.)

Such centralization of command headquarters and the attendant consolidation of information systems to monitor the events in their respective areas of responsibility may provide the foundation upon which to build a different organizational structure for the execution of battle plans. A virtual combat air staff concept may be a useful option in the future, instead of deploying a large physical staff to those areas of operation.

STUDY SCOPE, QUESTIONS, AND ASSUMPTIONS

The objective of this study was to investigate the nature of the future combat air staff in the context of air war in the information age. A number of significant questions shaped the inquiry and deserve attention: How did the current combat air staff organization evolve? What technological capabilities will yield opportunities to provide "virtual" links to military services and information once only available in theater? What can we learn about "virtual" organizations from the business world? What answers to the above questions may help define potential needs and develop specific future combat air staff requirement statements?

In addressing these questions, the major objective is to illustrate the substantive issues, rather than provide definitive solutions with specific manpower requirements and associated costs. From the authors' perspective, it is apparent that the current organizational centerpiece within the USAF is the composite wing. Despite monumental efforts to standardize objective flying wing organizations, each of the designated Air Force composite wings is unique, consisting of different aircraft types and deploying different numbers, depending on the contingency. No effort was made to create a notional composite wing for this study or even to use the composite wing as the standard organization that will deploy. All efforts to categorize job requirements were kept at the macro-level. The intent was to influence debate on the types of tasks required in theater or to be conducted through a communications link. The specific personnel numbers (presumably fewer), cost savings (if any), and job descriptions of the virtual combat air staff are best left to manpower experts once organizational structures and functions become clearer.

Some boundaries must be established to frame the debate. Our opinion, based on observing the changes outside the military world, is that the future combat air staff should be an organizational chameleon, changing composition and emphasis as the doctrine, strategy, and tactics of waging future air war evolve. The framework for study must encompass enough flexibility to address a number of diverse issues.

There was great temptation to consider the impact of joint service C^2 on future air operations and the future combat air staff. Many issues, including interservice politics, factor into a discussion of the Joint Task Force (JTF) organization. There is no doubt that the JTF model is the basic construct of the future deployed U.S. military force team. An attempt to cover all aspects of a JTF's operations is beyond the scope of this study. Some findings and discussions may be transferable to a JTF, but the emphasis remained on issues related to a deployed combat air staff.

Some underlying assumptions are also worth noting. The construction of a prospective combat air staff was undertaken with a focus upon future (2010–2020) potential needs and technological capabilities. No effort was made to justify current military force levels or deployment options. Advancing technology is likely to replace some current jobs and create others. Throughout the range of inquiry, the researchers assumed that manned aerospace power projection will remain the dominant method of waging air war during this time frame.

Although this work may shed some light on questions concerning future peacetime air base and air staff manning, that possibility is purely ancillary. The brunt of the work concerned investigating deployed combat air staff manpower functions and the myriad issues associated with "networking" capability and information to a theater air commander and his staff.

REPORT ORGANIZATION

This study seeks to shed some light on the questions of how virtuality in air combat operations may help meld the hierarchical and network organizational forms in the future. Chapter Two provides a brief history of the evolution of military staffs. Chapter Three gives an overview of the enabling technologies that are with us today and how those technologies may grow to shape our future. Chapter Four provides a review of the virtual organization in the corporate world, and Chapter Five explores some options and issues for applying those models to the U.S. military. In Chapter Six, we present a notional combat air staff scenario that is representative of the concepts and opportunities presented in the previous chapters. In Chapter Seven, our retrospect recaps the various issues raised and postulates a hybrid framework for applying virtuality to combat air staff operations.

Chapter Two

THE EVOLUTION OF MILITARY STAFFS

There are common themes, regardless of the historical application, that drive the requirement for a "staff" approach to waging war. Common, basic requirements that contribute to the need for a "staff" include command of troops (including morale), communication and coordination, operations and planning, intelligence, logistics, and the care of troops (supply, transportation, quarters, etc.).

These basic requirements forced ancient emperors, thought to have been direct descendants of gods, to depend on trusted followers as they went about waging war. These same needs shape current military organizations. There are, however, certain characteristics that distinguish a modern military staff system from the ancient staff model, including

- Methods to delegate authority from the commander to subordinates

- The supervision of the execution of orders

- A specified division of labor between different elements of the staff

- A structured educational program to train staff officers (Hittle, 1949, p. 9)

- Tools to support the staff in discharging its functions.

A brief discussion of this evolution is important to understand how military staffs have increased over the years in direct response to the increasing width, breadth, and depth of the battlespace. As we move into the information age, that battlespace extends to all regions of the world and into all five dimensions of warfare: air, land, sea, space, and cyberspace.[1] As a result, the historical evolution of staffs is more than likely not the correct path or the organizational response for future combat air staff structure.

DEFINING THE DIFFERENT VARIATIONS OF MILITARY STAFFS

In its simplest context, a staff's purpose is to help the commander with whatever the commander needs. Functions may include providing information, preparing details of plans, translating decisions into orders (and ensuring their execution), advising on relevant matters, and planning future operations. There are two broad types of staffs: general staffs and field staffs.

General Staff

The headquarters staff of a nation's military, or the military advisory group to the chief executive of the state, is the general staff. This element assists in the development of the overall defense strategy and policies for the country. The Joint Staff can be considered a general staff.

Field Staff

This element provides support for the combatant forces, also known as a unit or troop staff in tactical units, or "general staff with troops." The wing staff at a typical Air Force base (e.g., the 33rd Fighter Wing at Eglin AFB) is a field staff. (Combined Arms and Services Staff School, 1990, pp. 1–3.)

---

[1]Like information warfare, cyberspace has many definitions. Cyberspace can be defined simply as the global information infrastructure. A complex, and perhaps more esoteric, definition is the following: cyberspace "is a completely spatialized visualization of all information in global information processing systems, along pathways provided by present and future communications networks, enabling full copresence and interaction of multiple users, allowing input and output from and to the full human sensorium, permitting simulations of real and virtual realities, remote data collect and control through telepresence, and total integration and intercommunication with a full range of intelligent products and environments in real space." (Novack, 1991.)

Until the last century there was little to distinguish between a general staff and a field staff. One reason for this lack of differentiation is that, throughout much of history, the emperor, king, or monarch of a country at war also personally led his troops into battle. His general staff consequently became his field staff by default. As the complexity of war increased, a distinction emerged between a general staff and a field staff. This field staff (which we can think of as our deployed staff) evolved over several phases, from a smaller personal staff to a larger coordinating staff construct.

Personal Staff

In the Middle Ages, when combatants numbered in the hundreds or less, the commander directly controlled all troop actions. He could see the entire battlefield and give orders directly to his troops. Despite inherent complexities, the commander handled all aspects of the campaign. His assistants (personal staff) attended to immediate personal and subordinates' problems. The personal staff performed the functions of today's first sergeant, inspector general, judge advocate, and chaplain (along with some cooks and servants).

Special Staff

As the size of armies grew, the commander could no longer manage all the details of battle and the collective needs of his soldiers directly. Special staffs provided technical expertise in specific areas, allowing the commander to focus on the operational tasks of defeating the enemy. The special staff included support functions, such as provost marshal, public affairs, finance, supply, maintenance, administration, engineering, and topography. The special staff continued to report directly to the commander.

Coordinating Staff

As the number of combatants continued to grow, the dimensions of the battlefield became too extensive for an individual commander to monitor all actions of his troops. The weapons of war concurrently improved in range, reliability, and lethality. These developments forced the commander to rely upon assistants to help manage the tactical, as well as the support, functions of his army. Trained assistants emerged to coordinate different staff duties. The commander could then rely on this coordinating staff to provide the necessary information to make operational decisions and carry out orders.

PRINCIPAL PHASES IN THE EVOLUTION OF THE MILITARY STAFF

Military historians delineate three major phases in the development of the military staff. The nature of military staff employment in each stage provides useful trends for tracking the historical evolution of the military staff:

1. Ancient Empires (3000 B.C.–500 A.D.). Encompassing the period from the first signs of written history through the fall of the Roman Empire, the nature of warfare transitioned from unstructured brawls into recognized armies and the development of principles of war. Much of today's terminology can be traced back to the latter part of this time period.

2. The Middle Ages (500–1500). In the West, this time frame featured advances in weapon technology without a concurrent improvement in the organizational construct for waging war. Some countries improved upon concepts born during the preceding phase, although military organizational approaches appear basically regressive rather than progressive.

3. Modern Era (1500–present). This period began with Maurice of Nassau and King Gustavus Adolphus of Sweden, who many believe initiated modern military methodology and campaign development. Major advancements in both organizational philosophy and the art of war occurred during this most recent phase.

SUMMARIZING THE TRENDS IN STAFF DEVELOPMENT THROUGH HISTORY

In antiquity, many of the key developments in the evolution of the military staff were in response to the changing nature of the battlefield. Weapon technology advanced slowly, and there was a corresponding gradual increase in the size of armies and in fields of specialization. These developments made it more difficult for a single individual to take care of, control, or coordinate disparate activities effectively. The personal staffs that were established functioned as extensions of the commander, serving as his eyes and ears across extended distances. These staffs grew, and their individual compositions became more specialized to handle increasing complexity and diversity. The introduction of the concept of logistics brought an increased sense of permanence. Intelligence and reconnaissance functions grew in importance. However, even with the advent of specialization, the commander still retained direct personal control of operational and tactical planning and staffing functions.

After the fall of the Roman Empire, the next thousand years brought revolutionary changes in weapon technology but little change in staffing concepts and organization. Medieval armies were relatively small, and there was very little change in the speed at which information traveled or in the methods for gathering information. With a few notable exceptions, warfare remained a seasonal, impermanent, highly personal endeavor. The ethos of medieval warfare was the glorious knight in hand-to-hand combat with the medieval commander out in front leading his troops. While this approach might have contributed to maintaining troop morale, any sense of strategic military planning was lost. As such, military staff development may have included advancements in logistic and supply planning but little else.

During the 17th and 18th centuries, there were major technical and organizational improvements in military staff theory, but battles were still fought in much the same way as in ancient times. Commanders were warriors first, leading the troops into battle rather than directing from a headquarters. Prior to the 17th century, commanders were reluctant to split their forces because communication over long distances was problematic: There were few good roads, almost no good maps, and an absence of portable timepieces to coordinate attacks. (Van Creveld, 1985, p. 26.) However, toward the end of this period, the introduction of improved weapons, transportation, and communications technology resulted in a stunning growth in the size of the battlefield. Commanders, consequently, had to work from centralized positions in the rear, planning, plotting, communicating, and coordinating to maximize "presence" across a wide area.

This transition in the role of the military leader had a tremendous impact on the way staffs were designed and implemented to support the command function. Whereas medieval combatants relied upon courage and pure strength, the modern era has been characterized by carefully planned campaigns, more scientific use of weaponry, professional soldiers using well-developed tactics and strategy, and increasingly complex means to facilitate communications, command, and control. The increased importance of engineering and the sciences has required the development of specialists. The need for advance planning and orchestration of a growing variety of support functions has required the commander to depend more and more on his staff to perform the many specialized tasks that must be accomplished in fielding and directing a large force.

Technology has fueled much of the evolution of military staffs. Technological advancements have combined to expand the area of operations. This opening of the battlefield, while good for achieving offensive operational goals, has decreased a commander's ability to oversee the battle and provide direct orders. While the military's offensive capability has increased, the commander's sphere of influence has failed to keep pace. He therefore has grown more dependent on assistants to help manage and coordinate disparate activities.

ASSESSING THE IMPLICATIONS OF MILITARY STAFF EVOLUTION

The trends in military staff development point to continuous adaptations to respond to the need for maintaining and enhancing "situational awareness." The military unit must—among other things—understand its environment, have some knowledge of the enemy's position and intentions, be able to assess its own strengths and weaknesses, and know what other friendly forces are doing. In this regard, communication is key. Early long-distance communication systems were either visual (e.g., smoke signals) or via courier. The speed of communication was limited by the speed of a horse (about ten miles per hour) for centuries. Advances in communications have enabled more timely and complete information, which has contributed to enhanced situational awareness at all levels of command. With current technology, that "visual" sign can now be a video display, an air tasking order (ATO), a data link, or any number of other means for conveying information. Thus, spatial accessibility is no longer the limitation it once was.

Specialization has also proceeded beyond the individual level to that of the unit. Armies have evolved into divisions of combined arms. This increased complexity has brought many new requirements, among them the need for sophisticated war plans and detailed maps, increased intelligence availability, and a system of issuing documented (as opposed to oral) orders. (Irvine, 1938, p. 173.) Satisfying these many needs has called for a well-trained and disciplined staff, knowledgeable in a wide variety of specialties and conversant in both the science of current technology (mathematics, engineering, cartography, etc.) and the art of war (tactics and strategy).

The future promises combat staffs with the ability for almost unlimited access to command authorities and fielded forces via worldwide communications, increased situational awareness, and the ability to employ a wide range of technical expertise. Physical characteristics of command no longer dominate the debate on a commander's span of control. Future concerns revolve around efforts to manage incredibly large databases and information nodes, coordinate large and disparate forces, and at the same time provide only the most critical information at the right time to those people who must wage war in the information age.

It is evident that the evolution of military staffs has occurred as a response to the needs of the battlefield commander. The breadth and complexity of the staff has grown in parallel with the size of military campaigns and the responsibilities of those who manage their conduct. While staff size does not necessarily correspond with its added value, it is reasonable to posit that the accuracy and quality of information provided to the commander by his staff have been key to successful leadership in military operations. Thus, staffs will continue to be an instrumental part of any military organization. The question to be explored in subsequent chapters is how the staff will evolve in the future to take advantage of emerging technologies, to meet new challenges, and to respond to ever-changing external pressures.

Chapter Three

TECHNOLOGICAL CHANGE AND THE
VIRTUAL STAFF

In the realm of computers, networks, and all forms of communication, the pace of technological change is breathtaking. The strides being made in the information technologies are so dramatic that many observers now claim that we are in the midst of a revolution. Without debating whether or not such changes constitute a true "revolution," there is little disagreement that the future offers possibilities that were the stuff of science fiction only a few years ago. That the advancements now within sight hold the potential to radically change the conduct of warfare—and how it is run by military staffs—is not argued. What these changes will be like is argued. And yet, such a debate is good. We need to discuss and conceptualize these changes now, before opportunities pass us by.

Any projection for the future of technology is fraught with the difficulty of trying to differentiate a reasonable possibility from a pipe dream. Furthermore, how does one assess the "unforeseen"? Still, a look at current technologies and trends can help establish credibility for those who propose new environmental paradigms that govern how we look at the world and organize ourselves to deal with it. This chapter will attempt to do that. This is far from being an exhaustive survey, and it is hoped the few tidbits of information provided will serve to motivate deeper thinking on the subject and begin to galvanize the U.S. Air Force to anticipate how it will deal with such challenges.

In the world of military acquisition, reference is often made to the motivation for procurement of new systems as being "requirement pull" or "technology push." In a sense, the requirement-pull side of the equation is constantly there; users always desire better, faster, and more robust systems. Likewise, advanced technology often provides opportunities that did not exist before, making it a desirable commodity. To address its needs, the Air Force (and the DoD, for that matter) has a very formal system for identifying requirements and their sources. At this stage, no official requirements for future staff characteristics have been documented. One could thus fairly describe the current situation as one in which the concept of the "virtual staff" originates more from the technology-pull side of the equation. This state of affairs is, however, typical of the broader environment. As Vice Chairman of the Joint Chiefs of Staff (VCJCS) Admiral Owens said, "the basic rationale for defense planning has shifted from threat to capability and from liability to opportunity." (Owens, 1995, p. 35.)

Accordingly, while it is worthwhile to outline potential operational needs, none of this discussion should be construed as formulating a "requirement" that *must* be met. The purpose is to contemplate possible futures and drive out issues for consideration *before* they are overcome by events.

This chapter is divided into five major sections. The first section places the virtual staff within the appropriate military and technological dimensions. The next section takes a brief look at recent history in hopes of providing a basis for understanding the trends shaping technological development. This section is followed by an overview of potential operational needs and technical requirements as enunciated by some of our country's forward-thinking leaders and technologists. The fourth section will briefly review some of the technologies that constitute the underlying foundation for the systems of the future and will provide projections for their advancement. (The appendix found at the end of this report provides more detailed supporting material in this regard.) The final section will ruminate on the implications and issues such possibilities portend for the virtual staff of the future.

ADVANCED C⁴I

The virtual staff is most fundamentally understood in the context of advanced command, control, communications, computers, and intelligence (C⁴I). These are the primary functions and means by which military leaders and their staffs manage the application of force in a military campaign. Advanced C⁴I rests upon certain technologies to sift through data to extract and then transfer information:

> It is the system that converts the information derived from battlespace awareness into deeper knowledge and understanding of the battle space and involves everything from automated target recognition to an understanding of the opponent's operational scheme and the networks he relies on to pursue that scheme. It is also the realm in which target identification, assignment, and allocation take place. In sum, it converts the understanding of the battlespace into missions and assignments designed to alter, control, and dominate that space. (Owens, 1995, p. 38.)

To implement such a system in the battlefield of the future, military organizations will depend extensively on computers and data-gathering systems. Computers are appropriate devices to aid in this function, because they are exquisitely capable of accepting, storing, processing, and presenting information. Networks enhance the usefulness of computers by rapidly sorting and disseminating information among the machines with which they interconnect. Much like the interstate highway system, the network infrastructure must have several simple, but powerful, properties: It must be widely available; be easy to use; and most importantly, serve as the foundation for countless useful activities (Dertouzos, 1991, p. 65). If the implementation of such a system is successful, it will provide worldwide information access via common data via links expandable as needed to deliver the goods in the format best suited to user needs. Such a system "will provide all U.S. warfighters immediate access to the critical information they need, from any source, in any electronic form, to and from anywhere in the world." (Widnall and Fogleman, 1995.)

RECENT HISTORICAL EVENTS AS PORTENTS OF THE FUTURE

Two recent military operations may portend the effects of technological trends upon future combat air staff operations. The first example is the Gulf War of 1991. This conflict was marked by unprecedented use of advanced electronics and communications equipment to manage the war effort. For example,

- 118 mobile ground stations, supplemented by 12 commercial satellite terminals, were employed to handle satellite communications.

- Complex linkages tied U.S. databases and networks to the war zone and handled up to 700,000 telephone calls and 152,000 messages per day.

- More than 30 million telephone calls were made in the conduct of the air war alone (Roos, 1994, p. 31).

- At the height of the conflict, the automated message-information networks passed nearly 2 million data packets per day through gateways in the Southwest Asia theater of operations.

As stated by former Chairman of the Joint Chiefs of Staff, General Colin L. Powell:

> Personal computers were *force multipliers*. Efficient management of information increased the pace of combat operations, improved the decision-making process, and synchronized various combat capabilities. (Powell, 1992, p. 370.)

Nevertheless, Operation Desert Storm did expose key shortcomings in current technical capabilities. On more than a few occasions, this conflict served to illustrate the futility of gathering overwhelming quantities of data that could not be transformed into actionable intelligence, delivered to combat commanders where and when needed, and packaged in a form that allowed for immediate exploitation. True, staff support was hampered as much by organizational inefficiencies as by incompatible data links and databases, but the needs for information processing far outstripped the capability of the infrastructure to meet the demands placed on it. (Campen, 1994, pp. 17–18.)

A second example is the Haitian mission of 1994 to 1995, Operation Uphold Democracy. In Haiti, the XVIII Airborne Corps Commander, Lt Gen Henry H. Shelton, was provided an International Maritime Satellite suitcase-size communications station. Army Space Command also deployed three NASA Advanced Communication Technology Satellite terminals into Haiti, giving General Shelton secure video teleconferencing (VTC) capability with his commanders in Haiti; his deputy commander in Ft. Bragg, N.C.; Ft. Drum, N.Y.; and the Joint Chiefs of Staff in Washington (Garner, 1994, p. 22). Senior Air Force leaders later claimed this was the first time that the Joint Forces Commander, National Military Command Center, and Service Operation Centers had instant access to view a common tactical picture displaying everything from readiness data to imagery and weather. (Widnall and Fogleman, 1995.)

POTENTIAL OPERATIONAL NEEDS

It is anything but coincidence that the idea of the virtual staff dovetails extremely well with the future for C^4I systems and information warfare as envisioned by the top leadership across all the military services. Interestingly, this vision for the future is not devoid of a knowledge of the technical advancements necessary to make it a reality. Statements in this regard by military leaders may be broad in scope, but they also tend to be rather unequivocal and ambitious. Take, for example, this quote from General Powell:

> From the commander's perspective, information received should provide an accurate description of friendly, enemy, and neutral elements in an area of concern—the "battlespace." To provide the detail and quantity of information required, a distributed database needs to be created from information provided by all available sources. Intelligence, operational, logistical, and administrative information must be fused and distributed in such a way that it can be pulled from this global "infosphere" on demand. . . . This aggregate information should be analyzed to merge duplicated information into a single element. The expanding discipline of artificial intelligence (AI) gives great promise to help. The presentation of this information is also important. . . . Each [commander or logistician] should have the ability to shape the presentation of information as desired. Software-controlled customizing of each command node is the goal. . . . Future information needs on the battlefield will demand quantum leaps in processing power and memory capacity without increases in size and power consumption. Tomorrow's warfighter will require global access to information and transparent multilevel security in a laptop system. . . . The ultimate goal is simple: Give the battlefield commander access to all the information needed to win the war. And give it to him when he wants it, where he wants it, and how he wants it. (Powell, 1992, p. 370.)

The Air Force Scientific Advisory Board summarized the Joint Staff position with these words:

[T]he Joint Staff . . . is promulgating a vision for future warfare management entitled "C^{4}I for the Warrior (C4IFTW)." . . . The tenets supporting this vision are:

- 100 percent interoperability—ubiquitous connectivity and data entry only once in the system

- Common operating environment—standards to permit application portability

- Flexible, modular C^{4}I packages—to assemble compatible, interoperable C^{4}I capabilities

- Horizontal/vertical C^2—data, voice, video, or integrated information sent up, down, or laterally through the war fighting force or other organization; includes doctrine, standard terminology, and data to ensure common use and understanding

- Over-the-air updating—data bases automatically updated

- Warrior pull on demand—retrieve remote data from any place/any time

- Real-time decision aiding—supports adaptive planning

- Global resource management and control—any place, any time, any mission

- Adaptive safeguards—to assure protected, uninterrupted C^{4}I

- Seamless operations—into an interoperable, cohesive global network. (Druffel et al., 1994.)

This is a highly cogent vision for the future that is echoed by other military leaders. Take for example, VADM Richard C. Macke, previously director of the Joint Staff, who pointed to a global command and control system as *the* critical element to allow CINCs to configure their forces as needed. As conceived by Admiral Macke, such a system would "list readiness of available forces, along with their capabilities, so each commander could select from a menu of forces to configure the task force." He even goes so far as to state that this same system should be capable of drawing the organization chart, assigning specific units, specifying functions for each component commander, developing activation plans for reserve and guard forces, and providing alternatives for units not yet ready to join the fray (Ackerman, 1994, pp. 65–68). To this, RADM Charles R. Saffel, Jr., the Joint Staff deputy director for unified and specified command support for C^4, adds the ability to "look at the future battle arena— one hour, two hours, or even two days in advance of operations." (Ackerman, 1994b, pp. 68–70.)

Translating these operational needs into technology requirements is no small task, but it is doable. As shown above, General Powell recognizes that advancements will be required in artificial intelligence, software design, processing power, and memory. Other sources acknowledge that solving shortfalls in battlefield communications, especially intelligence collection, will only be realizable with exponential growth in signal-gathering and -processing capabilities. (Roos, 1994, p. 31.)

Air Force Lt Gen Albert Edmonds has said that "the satellite communications requirements of the U.S. Defense Information Systems Agency (DISA) are expected to grow almost five-fold between 1995 and 2010." The need to accomplish such a phenomenal growth rate has led the military to eye leading-edge technologies, such as direct broadcast satellite (DBS) services, personal communication services, and mobile satellite services. Edmonds further stated that DISA will require reliable transponder and gateway and transmission services (Doherty, 1995) and that transmission of huge amounts of data via broadcast transmissions will necessitate advancements in digital compression techniques. (Cooper and Holzer, 1995, p. 2.)

The litany goes on to include needed improvements in network management schemes (Ackerman, 1994b, p. 67), machine intelligence (Dertouzos, 1991, p. 64), and universal, seamless, access to data "for the naive, non-expert user" (Robinson, 1994, p. 25). Ultimately, the goal would be for the supporting technologies to "disappear" into the "fabric of everyday life" where they would become indistinguishable from it. "Ubiquitous computing" or "embodied virtuality" are the labels some have applied to such a "revolution" in information management. (Weiser, 1991, p. 94.)

In the final analysis, command staffs will require ever more information to retrieve, sort, digest, and disseminate knowledge so that the commander can make knowledge-based and objective-motivated decisions (as distinguished from knowledge generation and command implementation) with all pertinent information in hand and understood. In the past, as information-gathering capabilities outstripped the ability to absorb and digest, more available information has meant longer processing times and an increased danger of "failing to distinguish between the relevant and the irrelevant, the important and the unimportant, the reliable and the unreliable, the true and the false." (Newell, 1989, p. 26.) Advancing technology could allow information processing capabilities to catch up to the ability to collect data and bridge this critical gap.

TECHNOLOGY PROJECTIONS

Optimism that such a bridge can be built is founded upon observations of the progress that has occurred since the invention of the computer some 50 years ago. In the time since this seminal event, a phenomenal compounding of capabilities has occurred in many dimensions, including microchip processing speed, throughput, and sophistication. Advances have also occurred in such critical fields as display technologies, wireless systems, and security methods. Sustaining or even accelerating the progress seen to date may not be possible across the board, but that progress will continue is inarguable. Our purpose here is to provide the broad outlines of technological advancement even if the rates of change do not project consistently into the future.

Advances in processor performance will in the near term probably continue at the same pace, which has seen an order of magnitude increase in the number of instructions per second every seven years. Although physical limits that would preclude such a sustained pace are being approached, new techniques are being explored (innovative packaging schemes, opto-electronics, etc.) that hold promise to keep progress on track with recent historical trends. Bandwidth capabilities should also experience great strides from today's tens of megabits per second (Mbps) to tomorrow's tens of *giga*bits per second (Gbps) via optical fiber. Satellite communication links should lag optical fiber by perhaps only one to two orders of magnitude in the same time frame. Such throughput levels are within almost certain reach over the course of the next decade.

Advanced processing techniques will achieve many innovative and useful capabilities through use of such technologies as highly accurate character recognition, artificial intelligence (AI), "software agents," fuzzy logic, and neural networks. These technologies will enable users to gather, manipulate, manage, and understand huge amounts of data that would otherwise remain unavailable and/or incomprehensible. Visualization and presentation capabilities will also assist in these tasks by providing higher resolution, lower power usage, and reduced size and weight for all sorts of imaging devices from blackboards to helmet-mounted displays. Advances in the material sciences and software interfaces will also enable data presentation and comprehension like never before.

Mobility of computing and communication systems will also advance beyond today's cellular telephones to permit real-time video imagery transmitted by and received from individual soldiers on the battlefield. These transmissions will become part of a vast fusion of sensor data allowing construction of a complete picture of the theater of operations. Lightweight, plastic batteries; multicast algorithms; and adaptive network management schemes will be among the many technologies that make this sort of battlefield awareness possible. However, all may be for naught if data and communications remain vulnerable to spoofing, corruption, interception, or jamming. Thus, security technologies must of necessity advance greatly to address the many needs across the battlespace as the information dimension grows in breadth and importance. While the potential for technical advancement in this realm is difficult to assess, some technologies, such as encryption, will continue to remain viable, while others, such as advanced "firewalls," will rise to the fore as confidence is gained in their integrity and reliability.

CONCLUSIONS

Of all the progress in various technology disciplines, change in the infosphere holds perhaps the greatest promise to alter warfare as we know it fundamentally. Our senior military and civilian leaders recognize this and have embraced it as a necessary evolution from the days of the Cold War. The question is no longer "if," but "when" technology will advance to the point that it can be employed to give the commander the full measure of battlespace awareness these leaders have envisioned.

The dizzying advancements that we have witnessed of late that make the virtual staff a reasonable expectation and not a product of fantasy need to be placed within a greater context to be properly appreciated. It is easy to forget that modern information management, as an engineering discipline, is only about 35 years old. As one author has described it,

> we, as a technology community, are in the barbarism stage of development. . . . [H]istorians will look back at today as a primordial period. . . . [W]e must soberly anticipate that similar to most engineering disciplines, the technologies of the early stages are but an amateurish preview of what is to come. We must plan for incredible change now not only because we are confronted with a major transitional period, but because as information management and modernization matures, its rate of change will ever accelerate. (Boar, 1993, pp. 76–77.)

The survey of technology presented here (and developed further in the appendix) shows that the advances necessary to make the virtual staff a reality are either in work today or can be reasonably expected to come to fulfillment based upon historical trends. While technology will be a pacing factor, it does not appear that any intrinsic show stoppers lie in the path ahead. Realizing the potential of virtuality will probably rest on processing capabilities, certainly more so than on raw computing speed or memory capacity. It is not so much the storage or access to information that represents the ultimate challenge but whether or not the information can be properly digested for decisionmaking purposes within the time constraints allowed by the high operating tempo of the modern battlefield. It will be difficult to replicate the experience and judgment of the human element of the battle staff, which enables it to sort information and find that apparently insignificant piece of data that in reality has enormous implications for the conduct of a military campaign. Thus, formulating the virtual staff to take advantage of the strengths of both man and machine represents the ultimate challenge. To quote Andrew W. Marshall, Director for Net Assessments, Office of the Secretary of Defense,

> Those who make the right changes and use of technologies gain a quantum increase in capabilities .... [H]owever, ... the "fundamental problem is intellectual." Success lies not in the technologies themselves but in developing the right concepts of operations and organizational structures to best exploit them. (Morrocco, 1995, p. 22.)

A combination of hardware and software advances coupled with recent successes in isolated military operations is beginning to make believers out of skeptics who previously claimed that the fog and chaos of battle could never be adequately dissipated to permit highly automated and integrated decision processes to come to the fore. The potential is unquestioned, and experiments have begun to explore the envelope to determine just what the limits are and to identify areas where further development is required. Given the "open" nature of the enabling technologies, it is crucial that the U.S. military remain at the leading edge of efforts to optimize their employment and to construct the concept of operations that will dictate how the user will interface with them. Success in this endeavor is imperative: As General Fogleman, current USAF Chief of Staff, has observed,

> Those who capture this computing power and the corresponding speed of information flow are going to have a tremendous advantage .... History has taught us that if you can analyze, act and assess faster than your opponent you are going to win. (Morrocco, 1995, p. 23.)

Chapter Four

THE VIRTUAL ORGANIZATION IN THE CORPORATE WORLD

In the past, business organizational architecture mirrored the lines of the military hierarchical construct. Business organizations that conceived, manufactured, marketed, stored, and delivered products in the industrial age used many of the same hierarchical, centralized, bureaucratic organizational relationships characteristic of a large military organization. Information flowed up and down well-defined chains of responsibility, with limited lateral information transfer below the senior executive level. Middle-level executives performed bureaucratic functions that enabled the vertical flow of information without adding greatly to the value of the product. Predictable career progression meant job security.

Today, a revolution in information technology and telecommunications, combined with intense global competition, has made this industrial-age model increasingly obsolete. Large companies have drastically restructured, making traditional job security a thing of the past. Blue-collar jobs are disappearing, and white-collar jobs are becoming much less stable as the global economy moves toward a new equilibrium. In this new environment, a number of companies, both large and small, have had success implementing the concepts of a virtual organization. These concepts include the virtual (or temporary) company, the mobile workplace, global production networks, and organizational and cultural changes that make the virtual organization possible. With information technology also driving changes in warfare, perhaps the military could benefit from these trends in the business world.

THE VIRTUAL COMPANY

Although the rise of the virtual organization can be attributed to many factors, the simultaneous explosion of both computer and telecommunication technologies has been the most important. Increased processing power allows small, inexpensive, portable computers to handle more workload than the large, expensive mainframes of a generation ago. Similarly, massive quantities of data can now be stored (via hard drive or CD ROM) inexpensively in very compact spaces. Computer networks, becoming increasingly robust and user friendly, allow this data to be transferred across wide distances in near real time. For today's company, increased computing capability has been augmented by advances in communications, such as fax machines, cellular phones, and pagers. These factors, when taken as a whole, tend to mitigate the advantages of access associated with being located at the same site as the corporate headquarters.

A division of a large company, taking advantage of computer networks and advanced telecommunications, can operate just as effectively hundreds of miles away from the home office. In addition, the remote location might provide additional benefits, such as tax breaks, lower labor costs, or better interaction with the customer or client base. The idea that a large company can be divided into several divisions operating at different locations is not new. Companies have been subdividing and decentralizing various operations since the advent of the modern corporation at the end of the 19th century. Typically, such reorganizations have been attempts to gain a competitive advantage through increased efficiency. What has changed in recent years is the capability of individual divisions to work cooperatively from different locations, exchanging information on a daily basis. Both the quantity and quality of information that passes back and forth would have been unimaginable even 20 years ago.

If individual divisions of companies can now cooperate effectively from remote locations, can individual companies now work effectively together as well? The answer is yes, and an increasing number of companies are doing just that. By definition, a virtual company is a temporary network of companies assembled to exploit a specific opportunity. The sum of the individual companies acts as one *vir-*

tual company but with no hierarchy, central office, or organizational chart. Instead, the organization is held together by a complex web of individual contracts and obligations.

The move toward virtual companies has been the result of the emerging notion of "agility" and the efforts of individual companies to become agile. For a company, *agility* can be defined as the capability to operate profitably in a competitive environment of continually and unpredictably changing customer opportunities (Goldman, Nagel, and Preiss, 1995, p. 3). The notion of customer opportunity refers to a situation in which a market emerges to meet the new demands of a particular customer base. In the old mass-production–based economy, customers accepted a limited variety of goods and services in return for the lower costs made possible by economies of scale. Today, computerized manufacturing and information-intensive services have made possible a wider variety and customization of each product or service, while keeping prices as low, if not lower, than using mass-production methods. Innovation and technological advancement have led to a shorter shelf life for most products. This has allowed customers to *demand* more specific goods and services tailored to their specific needs. The value of a physical product is increasingly tied to its information content or its customer service advantages (e.g., a new computer bundled with hundreds of dollars worth of software applications or a new car sold with an extended warranty and roadside assistance). (Goldman, Nagel, and Preiss, 1995, p. 21.)

In this new environment, companies are forced to pursue new business opportunities rapidly, before customer preferences change or competitors bring an improved product or service to market. This competitive atmosphere forces companies to evaluate their core competencies and outsource operations that put them at a competitive disadvantage. For instance, if a new opportunity to produce and sell widgets arises, but a company cannot make one of the components, it will form a temporary alliance with a company that is better able to produce that component, *even if that company is a competitor*, so that they can both reap the profits associated with this unique market at this time. When the window of opportunity closes, and for whatever reason, the production of widgets is no longer profitable, the companies will disband their partnership and go in search of new opportunities. If the original company had instead decided to train its employees to make the widget component or had attempted to hire new employees to perform that function, it would not have been able to get its product to market quick enough or cheap enough to be competitive. Agility is the trait that allows companies to avoid this pitfall. The essence of agility is defining core competencies and continually looking for opportunities to use those competencies, in conjunction with other companies, to seize a competitive advantage in a constantly changing marketplace. The virtual company is the mechanism used to capitalize on these fleeting customer opportunities.

An example of this type of arrangement is the cooperation between Apple, IBM, and Motorola to develop and market the PowerPC computer chip. Since the mid-1980s, a common characteristic of the personal computer industry has been the competition between two competing platforms: Apple's Macintosh and clones based on the IBM Personal Computer (PC) architecture. Despite IBM's production of the original PC, the market has since been dominated by companies that manufacture less expensive PC clones. The biggest economic winners in the computer revolution have been Microsoft, the software company that has produced the operating system for all PCs and PC clones, and Intel, the hardware company that manufactures the vast majority of the central processors (or chips) that power them. Today, realizing that the most lucrative opportunities lie in the more profitable businesses of producing chips and operating system software, IBM joined forces with its former competitor, Apple, and its chip maker, Motorola. These three companies are working together to produce a common hardware platform, based on the powerful new PowerPC chip, that will allow customers to choose either Apple's Mac System 7.5 operating system or IBM's OS/2 operating system. IBM is willing to cooperate with Apple because it knows that only creating a large enough market for the new chip and hardware will make its share of operating system sales worth anything. A key to this strategy is the ability of all three companies to redevelop their hardware and software *concurrently* so that a common platform can emerge in the near future (Goldman, Nagel, and Preiss, 1995, p. 30–31).

Another, perhaps surprising, example of the concept of a temporary, virtual company is the Rolling Stones' *Voodoo Lounge* tour. The organization, run by tour director and promoter Michael Cohl, uses

250 full-time employees and may potentially produce worldwide revenues of $300 million. However, because the entire corporation will disband after the tour ends in 1996, it is actually a *virtual* corporation. The entire effort is a joint venture between Cohl, production director Joseph Rascoff, financial advisor Prince Rupert Lowenstein, CEO Mick Jagger, and the other members of the Rolling Stones. None of these partners are bound by an organizational chart, and, in fact, no central office exists. Cohl works out of hotel rooms, carting around a crate of files, a laptop computer, and a fax machine. Adopting the core principles of a virtual company, the Stones brought in Rascoff's RZO Productions to supply the stage, lighting, sound, and other elements of the show. The stage, which includes a 924-ft^2 video screen and a 92-ft lighting tower, requires 56 trucks to be transported from city to city. Since it takes four days to assemble, three complete stages leapfrog each other across the country. For overseas appearances, the company uses two 747s and a Russian military cargo plane. Rascoff uses a network of computers and fax machines to track costs, of which labor alone can be as much as $300,000 per city. The agility made possible by this organizational structure became evident when Cohl decided to add 23 unscheduled shows after observing the initial success of the tour. (Landler, 1994, p. 83–84.)

INFORMATION TECHNOLOGIES AND THE VIRTUAL ORGANIZATION

As mentioned previously, the concept of the virtual company is made possible by the enabling role of information technology, as well as organizational changes designed to take advantage of these opportunities. Perhaps the single biggest change, which marks the main departure from the mass-production era, is the ability to treat masses of customers as individuals (Goldman et al., 1995, p. 17). Computerized manufacturing allows companies to manufacture goods in smaller lot sizes with more specialization. Complementing this capability, companies can now analyze very discrete market data. Electronic commerce via credit card, combined with the data generated from bar code readers, gives companies a much clearer picture of what individuals are buying, when they are buying it, and what other products they buy in combination. This information allows companies to develop specialized products targeted at a niche market. As technology improves and individual consumers participate more in interactive computer networks, giving instant feedback, the size of the niche will actually become the individual customer. Computer processing power, automation, and increasingly intelligent software will allow companies to track the preferences of individuals and provide products customized to the individual's specifications. This degree of customization and support will be the critical element giving a company an advantage over its competitors. As mentioned previously, in more products and services in today's economy, information content is where value is added.

However, marketing to the ultimate niche comes at a cost. It may require changes in organization, management philosophy, and operations, as well as substantial investment in manufacturing technology, information processing, and communications networks. Another hidden cost is the increased training required to educate workers dependent on rapidly changing technologies. Ultimately, the benefits outweigh the costs when companies make the appropriate organizational and strategic changes that allow them to use new information technologies to take advantage of profitable opportunities.

One clear opportunity made available by technology is the advent of massive databases and client-server software, which now enable access at all levels and locations within an organization. No longer are field data stuck at remote offices and headquarters data unavailable to distant divisions. "Smart" organizations now enjoy the simultaneous benefit of centralization and decentralization (Hammer and Champy, 1993, p. 93). Field personnel can use information stored in a central database to operate more autonomously, making quick, informed decisions without having to call headquarters for approval. When these field personnel enter their own data into the centralized database, they give the headquarters-level leadership a clearer picture to aid decisionmaking, allowing for more efficient employment of company resources.

Another technological revolution is the new mobility of the workplace, made possible by portable computers and advanced telecommunications. The quantity and quality of information that can now be passed back and forth from remote locations has freed workers from the geographic constraints of the large, centralized office. Salesmen are more effective when spending time in the offices of clients instead of their own companies. In an environment where companies are trying to meet the needs of each individual customer, a presence at the customer's office gives invaluable insight into what could be done to enrich that particular customer. The mobile office can be a benefit even beyond sales. As shown earlier, the Rolling Stones' *Voodoo Lounge* tour was not tied to one geographic location; therefore, an office at a fixed site had no benefit. The people running the tour needed to be *with* the tour. Modern telecommunications and portable yet powerful laptop computers make this possible.

AMR Corporation, the parent of American Airlines, is another example of a company trying to use new technology to enrich the customer. AMR is developing a wireless communications network designed to allow passenger service agents to roam through airport terminals, seeking out customers to assist. The basis for this is a system that will let agents enter requests or data into laptop or hand-held computers, which will be linked via radio modem to the nearest cell site. The cell site will then be connected by landline to AMR's Saber Group core computers and their databases. These databases provide information on schedules, ticketing, reservations, seat assignments, gates, types of planes, baggage, weather, maintenance, etc. The actual information that can be pulled from this database or entered into it will be no different from what is currently available through desktop computers located at the ticket counter or gate. AMR expects the added benefit to come from increased customer service. AMR has spent $500,000 thus far developing the wireless network and expects each device to use $500 worth of wireless services a month. However, they are betting that improved customer service will more than offset that cost. For example, when a plane comes in late because of weather delays, the ticket counter is usually swamped with 20 or more irate passengers trying to secure new connecting flights. However, desktop terminals can only handle a fixed number of agents. With the new wireless system in place, a team of agents could be dispatched to the arriving gate, ready to arrange new connections and escort customers to their connecting flights. The next step will be to develop more rugged versions of the hand-held devices to be mounted on vehicles for baggage handlers and maintenance crews. The goal is to save time and reduce delays, which would go a long way toward paying for the new technology. Ultimately, freeing employees from their desktop terminals will allow the airline to make more efficient use of its entire workforce. (Dellecave, 1995, pp. 33–34.)

Another application of the mobile office is the increasing number of workers who telecommute. That is, they work at home or on the road, using a personal computer, interacting with the home office via telephone, modem, or fax machine. In 1991, American Express Travel Services launched a trial program called "Project Hearth," in which it provided its most reliable agents the equipment and telecommunications links necessary to do their travel counseling duties at home. Within two years, over 100 counselors in 15 locations had switched to home work. The program was such a huge success that American Express as a whole now expects 10 percent of its employees who do telephone order-entry work to work at home eventually. The typical work-at-home agent handles 26 percent more calls, resulting in a 46-percent increase in revenue from travel booking, which works out to about $30,000 annually. In addition, American Express saves on the 125 square feet of office space it allocates for each agent. Although Project Hearth started in Houston, it could provide substantial savings in New York, where American Express leases office space at $35 per square foot ($4,400 annually for each agent). (Sherman, 1993, pp. 24–28.)

A major feature of the new economy is the globalization of both markets and production, which has, in turn, brought about the need for global production networks. For example, a company, operating "virtually" to exploit a customer opportunity in a distant market, might use its own superior design capabilities while relying on other companies' local production facilities and distribution channels. A tool for this type of operation is Enterprise Integration Network (EINet), founded in 1994 by the Microelectronics and Computer Consortium (MCC) in Austin, Texas, with funding from the Defense Advanced Research Projects Agency (DARPA). EINet is an Internet-based database with the goal of providing access to information about the capabilities of hundreds of thousands of companies, including the cost and availability of their expertise and facilities and their terms for participating in collaborations. Emerging electronic commerce will be managed by Sprint. Other organizations are springing up to provide similar resources. They include

CommerceNet, based in Palo Alto, California; ECNet, developed by Arizona State University; and AgileWeb in Pennsylvania. (Goldman et al., 1995, pp. 27–28.)

In similar fashion, Ford Motor Company is committed to operating its own global production network. In 1993, Ford electronically merged its seven automotive design centers (Valencia, California; Dearborn, Michigan; Dunton, England; Cologne/Merkenich, Germany; Turin, Italy; Hiroshima, Japan; and Melbourne, Australia). Then, in 1994, Ford announced that it would merge all its activities in 30 different countries into a single global operation with vehicle development divided among five centers by vehicle type, not by national or regional market. The purpose of centralizing vehicle development is to cut development costs to allow Ford to produce a wider variety of cars, aimed at niche markets, in a shorter period of time. (Goldman et al., 1995, p. 28.)

"REENGINEERING" THE CORPORATION

An essential prerequisite for running a virtual organization is the implementation of an organizational structure and corporate culture consistent with the demands of the new economy. In what has come to be known as *reengineering*, organizations abandon old hierarchical arrangements for a flatter organization based on cross-functional teams (Hammer and Champy, 1993, p. 77). Using the power of modern computer processors and databases along with advanced telecommunications, companies have begun to empower individuals at all levels. With these tools, fewer managers can now oversee the operation of more employees. This capability has made obsolete the industrial-age model in which companies were divided into specialized divisions that were in turn coordinated by multiple levels of management.

In a reengineered company, divisions and departments based upon specialized functions are replaced by teams organized by process. Generalists, with access to databases and "expert software," do the work once done by specialists, as jobs change from simple tasks to multidimensional work (Hammer and Champy, 1993, p. 68). Case managers, with an eye on the entire process from start to finish, provide a single point of contact for clients and customers (Hammer and Champy, 1993, p. 62). In this system, time once spent providing checks, controls, and reconciliation between departments can now be redirected toward value-added work.

To benefit from the efficiencies of reengineered organizational structures, companies must also foster a new corporate culture and value structure. An empowered workforce must have the freedom to make autonomous decisions. Reengineering puts a premium on creativity, encouraging both managers and workers to think inductively, first finding solutions then looking for problems (or customer opportunities) to solve (Hammer and Champy, 1993, p. 84). Nurturing this type of employee forces companies to change how they hire, train, and compensate employees. Prospective employees must be evaluated on more than their education and skills. Intangible character traits, such as motivation, self-discipline, and desire to please the customer, are weighed just as heavily. Training shifts from an intensive regimen designed to teach one job to a continual educational process in which employees are kept abreast of technological trends. Compensation focuses more on performance measures and less on activity and seniority. Employees are encouraged, through bonuses and commissions, to creatively boost their productivity (Hammer and Champy, 1993, pp. 68–76).

Reengineering, therefore, changes nearly every aspect of a company. In summary, business processes are first identified. Divisions and departments are then replaced by jobs and structures designed around these processes. A streamlined management devises measurement and compensation systems that encourage the kind of values and culture necessary to support the business process. For example, if customer service is a key part of the process, employees will be compensated based on customer satisfaction surveys. If profit margin and competition are the primary concerns, employees will be rewarded for cost cutting. (Hammer and Champy, 1993, pp. 81–82.)

On the down side, nationwide corporate restructuring has displaced a significant number of workers. For the individual, it can mean a constant battle against obsolescence. Improved technology allows companies to use fewer, more highly educated, technically capable employees to do the work once done by a larger number of less skilled, more specialized workers. Even managers have felt the sting as reengineering has cut out many of the middle-management levels in the corporate hierarchy. A recent example of this kind of reduction is Mobil Corporation, which eliminated 4,700 jobs after posting impressive earnings in the first quarter of 1995. Mobil is not alone, however. DRI/McGraw Hill, an economic consultant, estimates that corporate profits for the American economy rose 13 percent in 1993 and 11 percent in 1994. At the same time, according to outplacement firm Challenger, Gray, & Christmas in Chicago, American corporations cut 516,069 jobs. Clearly, companies have learned to do more with fewer workers. Like commercial firms, military organizations must understand the trends, both positive and negative, before embarking on a reengineering effort. (Murray, 1995, p. A1.)

Chapter Five

ISSUES IN REENGINEERING THE U.S. MILITARY

As a result of the post–Cold War drawdown, the U.S. Armed Forces have dealt with reductions in personnel similar to those in the commercial sector. However, rather than use reengineering methods to eliminate unnecessary levels of management, the military has tended to keep the same hierarchical structure while reducing actual combat strength. Figure 5.1 illustrates the hierarchical changes associated with the typical corporate reengineering approach, while Figure 5.2 shows a similar reduction using the military's approach. In the past, the military purposely maintained a larger staff during peacetime to ensure the ability to absorb large numbers of untrained recruits and draftees during wartime. Today, however, the United States is more likely to deploy rapidly with its existing, highly trained force. The speed at which military operations develop, coupled with the complexity of modern weaponry, places the majority of wartime duties outside the capability of a new recruit in the current environment. Just as corporations make alliances with other companies rather than hiring new employees, the U.S. military will need to operate in a coalition environment, relying not just upon allied militaries, but also on the civilian world expanded to support wartime needs.

Military organizations may be handicapped by the tendency to restructure in small steps while retaining most of the existing organizational structure. Lessons from the corporate world emphasize that reengineering must be done companywide, *replacing* the old organization with a new one. The true savings come from the efficiencies associated with the new process-oriented approach, not from increasing productivity within the existing system. Another potential

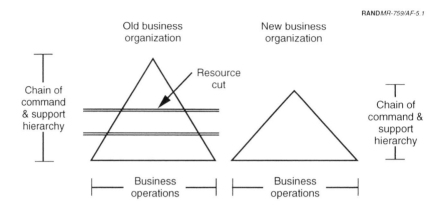

Figure 5.1—Business Reengineering—Business Operations the Same, but
Support Hierarchy Reduced

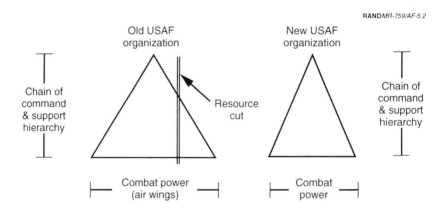

Figure 5.2—U.S. Air Force Resource Drawdown—Combat Power
Diminished, but Support Hierarchy the Same

problem is the free flow of information and individual autonomy
encouraged in a reengineered environment. A company can eliminate multiple levels of management, using pay incentives (a tool
unavailable to the military) to encourage relatively autonomous
employees to boost their productivity. The challenge the U.S. mili-

tary faces is developing new organizational structures and values that mirror the efficiencies and creativity businesses have gained in a reengineered environment, while at the same time retaining the elements of the old hierarchical, command-and-control, vertically oriented system (i.e., discipline, morale, tradition) essential for operations in a combat environment. This challenge gets to the very nature of the tension between hierarchical and network organizational constructs.

APPLYING TRENDS FROM THE COMMERCIAL SECTOR

On the surface, the Rolling Stones *Voodoo Lounge Tour* appears to have little relevance to challenges that face the U.S. Air Force (USAF). But careful inspection yields many striking similarities. The Air Force must be exceedingly agile, be prepared to dominate opponents across a wide spectrum of conflict, and provide support and services (from airlift to power projection) across a global stage. Frequent deployments require temporary coalitions and a constantly changing cast of players. Different contingencies will dictate the extent of support. Global communications and intelligence infrastructures can support a staff without regard to its physical position. Like the Rolling Stones organization, the Air Force effort is always "on the road" and must employ the same miniaturized information connectivity and processing power to allow a staff to coordinate a plethora of issues that include planning (adapting to changing opportunities), logistics (juggling resources), personnel (employee relations), and intelligence (market research).

Although the military operates in a different environment from the business world, the temporary, virtual organization may still have some relevance. Military organizations do not need to become agile to find new business opportunities. However, in many ways, the U.S. military needs agility more than ever to respond rapidly to crises and hot spots that flare up instantly in unexpected corners of the world. This requirement for speed, flexibility, and mobility rivals anything found in the business world. In addition, just as global competition is forcing companies to cut costs, the U.S. military must now operate within the constraints of a shrinking defense budget. Ultimately, the U.S. military is competing with the militaries of other nations for supremacy on the battlefield. However, the lessons of the individual company, operating in a larger virtual organization, using the concept of agility to define its core competencies and outsource its inefficient operations, could be very beneficial to the U.S. military as it operates in the environment of coalition warfare and peacekeeping. Also, the individual services and agencies within the Department of Defense need to evaluate their core competencies to form a more efficient cooperative arrangement. To go one step further, the Air Force, as well as the other services, might benefit from a similar examination of their internal agencies and commands. Using this broad view, the business world may have some innovative ideas that the military could find useful.

As examples in Chapter Four illustrate, the need for agility in the corporate realm requires quicker, decentralized information dissemination. This quickening pace might doom hierarchical models in the future information-driven approach to projecting military force.

THE CHALLENGES TO REENGINEERING

With an understanding of the historical trends of military staffs and examples of corporate adjustments to an information-age environment, it is now possible to discuss some of the issues raised by the creation and exercise of a virtual-staff approach for management of an air campaign. At this stage, our discussions are mere speculation about the possibilities and the pros and cons of such an approach, given the current lack of experience, other than in some initial wargaming simulations with relatively modest experimental C^4I systems.

Among the many issues facing the employment of the virtual staff concept are the following:

- How should the staff be organized to execute its tasks? (Models include the hierarchy, the network, a horizontally flat structure with an extended span of control, a vertically integrated structure, and hybrids.)

- How is "command" exercised in a decentralized system?

- What is the appropriate rank structure among those on the immediate (deployed) staff as opposed to those supporting via virtual links?

- What incentives must be in place to ensure virtual staff members support the deployed staff as well as they would if they were part of the latter?

- How might operations instructions, such as the ATO or its successor, be developed and disseminated by the virtual staff?

- What are the requirements for training of staff personnel to operate in the virtual environment?

- What are the issues behind "generalists versus specialists" considerations when manning a virtual staff?

- How will the virtual staff operate in the face of determined jamming of links and attacks on the integrity of databases?

- What is the best way to structure a virtual staff approach in the joint and coalition warfare arenas?

- How will the interface with the public be handled when the explosion of information availability enables civilians to gather data of intelligence value (similar to that procured by military systems) and then to interpret it independently of military analysts?

Obviously, these and other issues will not be definitively resolved here. By highlighting these pertinent issues against the backdrop of the preceding chapters, one may begin to conceptualize the critical characteristics of a future virtual staff. The discussion could, as well, ultimately guide us in learning what a proposed future virtual air staff should *not* be. However, in the next few pages, we will briefly examine areas that must be considered in a discussion about the application of this virtual combat air staff concept. The areas that we address are command considerations; operations planning and management; direct staff support, planning, and management; training needs; information compromise and denial; application to joint and coalition operations; and potential interactions with the public.

Command Considerations

In the past, staff role execution necessitated being in close proximity with all other supporting staff elements to coordinate action. Staff virtuality represents the possibility that technology may soon permit staffs to go about their functions without all elements being in the same physical location. Should this potential become reality, staffs could become not only more efficient and streamlined but could produce even higher-quality output.

Consider, as an example, an air operation in the Middle East like Desert Storm or in Bosnia like Deny Flight. The number of staffs and staff personnel necessary to support such operations under the existing paradigm could include (but would not be limited to) CENTAF, CENTCOM, USAFE, ACC, numbered air forces, elements of the Army and Navy, and the JCS. Each of these organizations employs staff intelligence experts with responsibilities for detailed knowledge about the opposition's air force. Currently, a significant number of people on each of these staffs do very similar jobs, with their own individual databases and information sources. Consider that each of the major intelligence agencies that supports military operations employs a similar staff with similar responsibilities. What if the staff support function included only *one* enemy air force intelligence cell, under a single lead, composed of individuals from supporting organizations, all linked virtually? There might also be occasions when experts from outside the traditional intelligence community (e.g., labs, operational test ranges, contractor facilities) could contribute substantially to the analysis of enemy capabilities and intentions.

Adoption of the virtual approach could, conceivably, greatly streamline staff requirements (i.e., reduce the absolute total number of personnel) and ensure all players had access to the same, more extensive, information base. Greater accuracy of intelligence assessments could result from the synergism brought about by personnel with differing perspectives looking at the same problems and with shared responsibility to populate and maintain currency of a common information source. The circle of "contacts" brought to the group by each staff member would, in sum, be widened considerably. On the other hand, there could be a loss of valuable redundancy and/or healthy competition in views, because staffs will have been reduced everywhere through a form of televideo consolidation. Such dependence on a single communications link to a single individual may bring additional vulnerabilities as yet unconsidered. However, in assessing the utility of this virtual concept, consideration of adequate protection for communications lines and retaining access to a robust communications structure will be needed. Along the same lines, if those automated tools and robust communications are severed or disrupted, the implications of having a very small combat air staff forward may necessitate an entirely different approach to continuing the air campaign over what is done manually today with a large, forward deployed air operations center staff. Balancing the concept of a single individual deployed forward and virtually linked against the current manpower-intensive concept will be necessary when debating the future application of virtual concepts while leveraging advanced information tools.

While such considerations point to the great promise held out by the virtual staff, the concept is not without potential pitfalls. Some of the questions regarding its implementation were raised above. A natural question is to ask: To what level will the commander direct queries or action items if the entire warfighting organization is literally only a "click" of the computer mouse away? If the point of expertise for a particular operational or support function is not resident with the commander, how will such an expert be tasked to provide staff support? Can a virtual command structure be implemented to override peacetime relationships so as to ensure that the warfighting commander gets timely staff support from units not his own?

Another issue has to do with the exercise of the command function. Networked relationships imply a more decentralized command-and-control structure both within and between the military services. Despite such decentralization, it may become possible—indeed tempting—for higher military headquarters to exercise greater control over operations (i.e., to micromanage) via information technologies. This paradox comes about because the new technologies make possible a greater "topsight" of the battlespace (Grier, 1995, p. 37). Just how close a commander gets to his operational forces or supporting staff may pivot more on personality than on any institutional impetus. It is perhaps unfair to say that the technology will necessarily lead to greater micromanagement, but it is certainly not beyond reasonable expectation. It may be prudent to implement organizational impediments within the virtual staff structure to discourage such tendencies.

One could also ask what sort of rank structure should exist in a networked system to permit the free flow of ideas yet avoid the problems inherent in work by "committee." Studies have shown that data networks induce participants to communicate more frankly and equally and to generate more ideas than is the norm. A significant contributor to this attitudinal shift is the lack of social cues, such as appearance and voice inflection. However, this greater democratization is also associated with increased decisionmaking times (Sproull and Kiesler, 1991, p. 119). The military, being a hierarchical organization with rank a traditional trapping reinforced by uniform wear and title, tends to magnify the effects of social intimidation. VTC technologies might only perpetuate such trappings and do little to shorten decision time cycles. Depending on the makeup of the virtual staff, one can foresee higher staff echelons employing officers junior to equivalent staff positions at lower command levels. Which officer should then be the one to report to the commander, in particular, when few of the staff officers are collocated with him?

A subtle issue is one regarding the psychological effects of the new information technologies on the exercise of command. As brought out in Chapter Two, on the historical development of staffs, such organizations have typically accompanied the operational warfighting forces to the campaign theater. It is worthwhile to ask how the employment of virtual organizations will affect their judgments and decisionmaking processes. For example, might not a virtual organization be more likely to send troops on risky operations because it does not share at least some measure of the same danger, that is, find itself within the range of enemy fire? The opposite issue is also of concern insofar as it is conceivable that some personnel in higher echelons might lose that certain measure of distance that allows them to make the hard decisions that mean sending men and women to their deaths. If practically "being there" becomes possible through telepresence technology, can virtual staffs continue making sound professional judgments when the results of their decisions are instantly and graphically displayed before their eyes? A lieutenant lacking calmness under fire might get his squad killed by his hesitation, but the poor judgments of a staffer might lead to more widespread casualties. In a similar vein, might instant feedback distract virtual staffers from taking the long view, insofar as they should plan for winning the campaign and be less concerned with achieving victory in each and every constituent battle? Certainly, these and other issues are faced by traditional staffs, but virtuality may magnify the effects and possibly reduce some of the checks and balances that serve to mitigate poor staff work.

Operations

Operational planning and management of an air campaign could very well benefit from employment of virtual organizing principles and the parallel supporting technologies. One might even go so far as to say that, in light of the very nature of the mission, advanced information technologies will be increasingly employed (to satisfy the demands for data flows and communications needs), and their very presence will lead to a natural migration toward a virtual paradigm. However, an organizational network that has been designed from the beginning to meet the demands to be placed on it, will undoubtedly serve better than one kluged together because of a lack of foresight.

It is perhaps an understatement to say that air operations is an area in which staffers have a number of responsibilities whose execution will be made more efficient through access to advanced information tools. From conceptualization of the overall strategic plan to implementation of real-time mission modifications, ever greater amounts of information need to be gathered, filtered, digested, and reformulated into a usable plan to do something. This "something" can be as overarching as the strategic plan of attack to bring about a successful conclusion to the war or as specific as the daily ATO. For the purposes of this discussion, these two extremes will serve as points of departure for speculation on the application of the virtual construct to the operations staff.

An interesting thought experiment is to speculate on how the development of the strategic air campaign for the Gulf War might have proceeded differently if it had happened in a virtual environment. The genesis of this plan, known as "Instant Thunder" in its initial formulation, has been documented elsewhere thoroughly and has animated debate about the role of airpower in modern warfare.[1] To

---

[1] For a good historical treatment of the development of the Instant Thunder plan and related airpower debates, see the two-volume set consisting of *Heart of the Storm* (Reynolds, 1995) and *Thunder and Lightning* (Mann, 1995).

begin with, it should be noted that much of the development of the Instant Thunder plan is attributable to the efforts and personality of Colonel John A. Warden III, along with his colleagues in the Directorate for Plans at the Air Staff. The plan was developed in an extremely compressed time frame using the expertise found in people, documents, and databases for the most part resident in the Washington area (Pentagon, Bolling AFB, CIA Headquarters, etc.). However, inputs to a much lesser degree were also received from organizations elsewhere, such as Headquarters Tactical Air Command (Langley AFB, Virginia, now Headquarters Air Combat Command [ACC]), Headquarters CENTCOM (MacDill AFB, Florida), and Headquarters 9th Air Force (Shaw AFB, South Carolina). For the most part, these inputs were received through one-on-one contacts made by personnel traveling from one location to meet with people at another. A lot of running around and hard work was required to gather data on the aircraft, weapons, targets, logistics flows, etc., necessary to execute the kind of strategic air campaign that Instant Thunder represented. Frequently, those who opposed Warden's ideas objected on the grounds that certain details were left unaddressed. Other times the objection was raised that Warden and his group had usurped the responsibility of the in-theater operations staff in formulating their proposal.

The thought experiment begins by postulating that much of the data gathering could have been greatly eased by having it available through networked databases and communications links. Target and threat locations, aircraft weapon requirements and capabilities, unit availabilities, logistics constraints, and more could have been quickly collected, studied, traded, and finally woven into workable alternatives with enough detail to give decisionmakers the necessary insight to determine optimal courses of action. Much of the concern about staff roles and responsibilities could have been muted through application of a network structure using virtual connections to link the appropriate staff elements together. Objections could have been raised sooner to permit counterarguments or revisions to be formulated as early in the planning process as possible. One result would have been a great savings in time—time that could have been spent running simulations or excursions or developing detailed alternatives, the merits of which could be weighed against each other. Another result would have been insight—insight into the options available, the reactions and thinking of the leadership, and the roles to be played. Still another result would have been greater confidence—confidence that the plans generated could be executed as envisioned and that they would achieve the anticipated results.

All this is not to say that there would be no problems. Personalities could still play a strong role, especially to the extent that rank could be used as a bullying tool rather than as the medium through which guidance is proffered. Assignment of lead responsibilities might be delayed in the chaotic atmosphere typical at the beginning of any crisis. Participants might place too much faith in the mountain of information available and in their analysis tools, forgetting for the moment the inescapable uncertainties inherent in the data and the limitations of the tools with which they are working. This tendency to implicitly trust all of this new technology may become more prevalent as the U.S. Armed Forces are populated with men and women who have grown up with advanced technologies and virtual worlds a part of daily life. As such, training will need to address the concerns unique to virtuality and deal with these, and other similar, changes in the world.

At the other end of the spectrum is the complex and time-critical ATO. For air campaign planners, the ATO represents

> the principle instrument for allocating the use of air power in the theater, translating the MAP [master attack plan] into specific wing tasking and instructions. (Winnefeld, Niblack, and Johnson, 1994, p. 136.)

During the Gulf War, the ATO—which was issued daily—was a voluminous document (frequently 300 pages, but as high as 900 when printed) produced using an obsolete system known as the Computer Aided Force Management System (CAFMS). CAFMS consisted of mainframe computers at the Tactical Air Control Center (TACC) with remote-receive SHF terminals located at wing-level units. Since Navy carrier air wings did not have such terminals aboard their ships, the ATO was delivered in hard copy by courier. Air Force B-52 wings also did not possess CAFMS. The ATO was produced on a 40+ hour time cycle, such that three ATOs were in existence at any one time—the one being executed, another being finalized and transmitted to units for execution the following day, and a third in the initial stages of preparation. The ATO assigned aircraft to targets; provided instructions for airspace coordination, communication frequencies, and call signs; gave combat search-and-rescue operation procedures; listed targets precluded from attack; and tasked noncombat and support missions, such as airlift and in-flight refueling. The ATO essentially provided oversight to air operations involving on the order of 2,500 daily sorties spread over more than 90,000 mi^2. The difficulties experienced with this system forced many Gulf War airmen to create workarounds[2] and to "game" the system to get the job done (Winnefeld, Niblack, and Johnson, 1994, pp. 137–139). Interestingly, many of these workarounds involved greater ad hoc use of information technologies and network forms of communication. Yet despite its shortcomings, the Gulf War–era ATO represented a clear evolution in the direction of virtual staff operations. It employed staff personnel from widely disparate organizations and depended on network links for the dissemination of information. While greatly centralized in its continuous development and issue, it demonstrated the principles of decentralized execution and the need for agility to adapt to rapidly changing circumstances.

Clearly, such a mechanism as the ATO could benefit from the technological advances associated with virtual principles. A future ATO system, such as that represented by the evolving Contingency Theater Air Planning System (CTAPS) or the future Theater Battle Management C^4I (TBMC^{4}I) system, may ideally be much more highly automated, perhaps employing optimization techniques, so that minimal effort is expended to achieve maximum effect. Such techniques would sort through all the demands placed on the ATO to execute the MAP as constrained by the limitations of assets, threats, and logistics. The ATO would be compiled much more rapidly, to take advantage of the latest information on the status of the warfighting infrastructure and intelligence about the enemy. The more fully integrated it would be with other information sources, the better it would perform this function. For example, imagine an ATO-generation system coupled to a battle damage assessment (BDA)

---

[2]This was done using existing communication networks and information technology tools. As such, "workarounds" were more commonplace in a combat situation than previously understood, further underscoring the need for some sort of formalized adoption of those workarounds into a "business as usual" construct.

database. As imagery, signals, and human intelligence information from all collection sources flowed into the database about the condition of targets attacked, that information could be used to update or reprioritize target listings. Extending the example further, imagine how much better the ATO could be constructed if it were tied into logistics and maintenance records that provided up-to-date information on weapon expenditures and aircraft availability. Integration of staffers at each affected location could also allow them to participate in the planning process of an ATO that would take advantage of each individual unit's strengths and compensate for its weaknesses.

Ultimately, one can even imagine an ATO that enables real-time adaptive mission assignment and execution. Aircraft might take off with generic or specialized payload mixes, but without specific orders, only to receive them in flight while loitering or heading toward the general vicinity of a target set. Mission video could be downlinked real-time to enable BDA and generation of additional orders for immediate restrike of high-value, time-critical targets such as tactical ballistic missile (TBM) launchers. In this case, virtuality will have extended into the cockpit and will potentially have eliminated inefficiencies associated with the involvement of intervening staff levels.

Direct Staff Support, Planning, and Management

The ability to serve a commander's information needs and support deployed troops with a virtual rather than physical presence is well within current technological possibilities. To provide a current real-world example, the authors of this report work at a location outside of a traditional air force base. Like a multitude of personnel assigned to Air Force detachments worldwide, there is no physical support staff on site. All administrative support, personnel actions, and travel and military pay functions are handled electronically or by mail. In the area of assignments, very few Air Force personnel physically work assignment actions from the Air Force Personnel Center. Very few physically deliver messages to the base communications center. Nobody receives a government check in person at the end of the month. Such trends will continue.

The unit cartographers, once-numerous members of a combat staff, now adequately handle the military's requirements for maps from a centralized organization, the Defense Mapping Agency. This agency is currently in the process of merging with the Central Imagery Office to form a National Imagery and Mapping Agency. Formerly critical members of a combat staff, such as messengers and signal officers, have been replaced by technology. Yet, the adoption and employment of such advances may be more revolutionary than the technologies themselves. Moltke's success during the Franco-Prussian War of 1870 was due to his revolutionary use of the existing transportation technology. Prussia was not the only European country with a railroad system. Moltke's organization of the Prussian Army capitalized on the railroad's speed and predictability to beat a numerically superior French force. Coupled with the use of the telegraph, the Prussian Army could deploy and coordinate separate force actions in a superior manner.

Still, the concept of "face time," that is, time spent in the physical presence of the commander, will probably continue to have significance. Face time has traditionally been viewed as critical to the combat support staff officer regardless of his expertise in administration, intelligence, personnel or manpower, finance, logistics plans, or other fields. Since the commander is, and in all likelihood will continue to be, an operator, face time allows the staffer to provide the commander with personal insights which allow him to appreciate and capitalize upon information properly outside of his own individual field of expertise.

This observation can be taken one step further. While it is important to take proper account of technological advances, the notion of "context" should remain significant in supporting staff efforts. In-theater personnel, such as contracting officers (to interact with host-nation contractors), supply specialists (to monitor and maintain logistical support), and deployed computer and communications specialists are likely to remain critical in the future. It may be only the staff oversight of these functions that, to an increasing extent, takes place virtually.

Training the Members of a Virtual Combat Air Staff

The importance of proper and frequent training to attain high-quality performance capabilities is self-evident. Staff operations in the virtual environment are no different in terms of the need to "train as one would fight." Since air campaign planning and execution in the virtual staff context have not been performed before to the level proposed here, training and simulation will be of critical importance to ensure optimal implementation of the concept. Once a working construct is established, continued experimentation and training will be required to allow for improvement and maintenance of currency.

Besides having a functional expertise, such as intelligence or logistics, the virtual staffer will of necessity be someone with an in-depth knowledge of computer systems, including networking schemes, software applications, and means to ensure security. Expertise in database management and configuration control will be a valuable skill, as will a background in telecommunication technologies. Proficiencies such as these imply years of training and development. To the extent that the technologies can advance to permit expertlike skill in operation by nonexpert personnel ("user friendliness"), training requirements can be relaxed. Experience has shown, however, that the greater the technical background, the more likely the user is going to be productive in working with complex systems. Furthermore, with the current trend toward procurement of commercially developed systems that incorporate the latest technologies, training programs will have to adapt to the rapid pace at which such technologies are introduced into the marketplace. Likewise, security training to protect this investment and these now more-vital systems will be required to ensure that adversaries do not exploit the widely known faults of various commercial products and services. Embedded security awareness and an understanding of potential problems may be one course of action similar to that of existing programs.

The question of generalist versus specialist training persists. In many cases, virtual links to supporting organizations will allow greater access to specialists and specific databases. The virtual combat air staff member will have to balance a generalist approach, so that he will know what sources to access for specific tailored information, with a specialist background that allows proper understanding and use of each critical piece of information.

All of these considerations have profound implications for how the military fosters the development of needed skills and identifies personnel for assignment as virtual staffers. It is not inconceivable that proficiency requirements may necessitate constant familiarization and education to allow the combat air staff member to stay abreast of all the pertinent developments and applications.

Information Compromise and Denial

A totally integrated C^4I system is a prime target for information compromise, corruption, spoofing, denial, and destruction—the domain of "information warfare." A total dependence on these technologies creates the potential for a catastrophic single-point failure or security breach. A high dependence on communication nodes will almost certainly invite attack by adversaries. It might prove to be a relatively straightforward task to overload communication links with extraneous or erroneous information and oversaturate the system. At a minimum, longer processing times would result in the loss of near–real-time advantages that technology advances promise. Or the integrity of data may be corrupted to a level that users no longer "trust" the system and will disregard its support. With an increased use of the broadcast spectrum, there is a corresponding increase in the opportunity for interception. Communication links have been exploited for centuries, and knowledge of an opponent's intentions before they are executed or as they are happening can, among other things, negate his use of surprise—one of the basic principles of war.

If deployed staffs are to take advantage of virtual concepts, it would seem wise to implement measures that enable quick reconstitution should vital links be interrupted or databases and application tools be corrupted. Disruption need not be total for unacceptable levels of degradation to be experienced. For example, it may take no more than the *late* arrival of critical guidance to operational forces for a campaign to be severely disrupted. It should also be recognized that measures to counter jamming are often accomplished at the expense of throughput. Jamming attacks could turn a VTC into something less than a telephone conference call. If the intent had been to provide a briefing using slides or a combat video, the briefer would now be faced with the problem of quickly finding other means to convey the information.

Jamming and information-security issues have been adequately addressed in the past, although our modern systems have never been tested by a determined and sophisticated adversary under wartime conditions. It seems reasonable that technology will continue to advance in areas that will work to provide secure, anti-jam communications, but the confidence in such measures needs to be weighed against the benefits that accrue from redundancy. Indeed, the whole concept of operations must be scrutinized to identify vulnerabilities and to conduct trades to see if the advantages to be gained from adoption of virtual operations outweigh the risks. It must also be scrutinized to ensure that the implementation of virtual operations is approached with full knowledge and understanding of the risks, providing a robust, on-demand communications structure to support the deployed staff.

Joint and Coalition Virtuality

The advances in information technologies could provide great assistance in the conduct of joint and coalition warfare. In a world of networked resources, service and national boundaries could soon become little more than sieves through which data and communications pass to allow warfighters to get on with the job. Imagine, if you will, how virtual staffers might contact their counterparts in other services or national forces, exchange information, and develop plans, all without having to go through the bureaucratic obstacles inherent in hierarchical structures. Language differences could be overcome through on-line, real-time translation. Intelligence information gathered by assets under the control of disparate forces could become instantly available to an entire JTF or to all forces in a coalition. Assignment of liaison officers to centralized headquarters might no longer be necessary. In sum, many of the traditional sources of friction inherent in such relationships could be eased if not eliminated by future technologies and employment of virtual principles.

Despite this promise, the record for efficient communications between joint, let alone coalition, forces has not been good. Most are now familiar with the anecdotal tale that, during the invasion of Grenada, a ground soldier had to telephone the United States, using his personal credit card, to request a fire-support mission near his location. Ground forces in this operation had equipment that was not compatible with that of the air units assigned to provide close air support. Such problems in communication and coordination with international troops are sure to be a much greater challenge, and no amount of technological advancement will overcome lack of agreement to common standards. Opportunities for fratricide will multiply on the battlefields of the future if for no other reason than the growing number of different identification friend or foe systems.[3]

It would also seem that many of the other concerns raised above apply just as acutely to the joint and coalition scenarios. While it may be clear who the JTF or supreme allied commander is by way of political edict, how the (virtual) staff should be formulated to support him is much less certain. In a networked system, all players may want equal billing when their relative status (associated with the parent organization) becomes less differentiated by the computer screen. In the Gulf War, great pains were taken to make visible the coalition nature of the decision processes used in the planning and prosecution of the campaign. Will it still be possible or desirable to achieve such visibility when the coalition staff does not even reside in the same building? And how does the commander deal with a "subordinate" foreign military leader who has access to all the same information as his staff and chooses to interpret it differently in public?

Interaction with the Public

The potential for greater public visibility into the conduct of war raises its own set of unique issues. Using a simplistic approach, two levels of societal interactions can be identified to include those with the media, as well as those with individuals having public (and perhaps surreptitious) access to network information sources, such as the Internet. However, these two levels should not necessarily be thought of as completely distinct. For example, during the Gulf War, news organizations, such as CNN, were major information sources for our military leaders, as well as for the public. The advances in commercial telecommunication technologies now allow just about

---

[3]A program to help reduce this problem, the Mark XV IFF program, was canceled in the mid-1980s because the different countries were unable to adhere to a particular standard. There were also developmental problems. Even within the North Atlantic Treaty Organization (NATO), there are problems when countries have to interface equipment despite the existence of NATO Standards.

anyone to become an information source or customer for both data and imagery. This information can be transmitted to a great many people, who in turn are able to receive and further disseminate it.

Could military public affairs and civilian news organizations combine operations virtually—sharing the same information bases and cooperating in their responsibility to inform the public? Is it out of the realm of possibility that public sources could be incorporated by (virtual) operators and staffers alike to enhance the performance of their respective missions? Will war correspondents and photographers be allowed to operate throughout a theater (including near the front lines), or could they be replaced by commercially run, unmanned airborne reconnaissance vehicles? What are the issues associated with the military and media "sharing" the same information on the progress of operations and locations of enemy troops?[4] Will the military have to compete with the news media for communication frequencies and bandwidth?

Once the technology is available to empower individuals to gather information that was at one time available only from the media or via press conferences, the military will have to scrutinize its information availability and dissemination means even more stringently. Information availability on this order presents potentially overwhelming security problems for the military and its conduct of successful combat operations. It also puts an onus on the media to be more responsible in its collection and dissemination of information, to prevent the compromise of data otherwise unavailable and of value to the adversary. Given the continuing decentralization of information sources and the independence characteristic of those working through networked systems, such control will be difficult to exercise.

---

[4]In 1989, the Carnegie Endowment for International Peace conducted a study that showed that commercially available satellite imagery could be used in the identification of military facilities or capabilities. Using imagery obtained from the SPOT or Landsat satellites, radars, supply dumps, major headquarters facilities, airfields, aircraft, rockets and artillery, missile sites, surface ships, and surfaced submarines could be identified. (Bankes, 1992, p. 17.)

CHALLENGES AND OPPORTUNITIES

To this point, the possibilities for using new organizational forms and virtual connectivity have been based on an exploration via conceptual thinking. The rationale for our discussion has been found in the need for the U.S. Air Force to continue to perform a multitude of operations across the conflict spectrum in a period of overall military force reduction.   As a result, fresh approaches to communicating and formulating organizational structures to be able to respond to these operations appear warranted.

The temporal aspect of the virtual combat air staff includes combinations of force mixes that come together for a short period of time to accomplish a specific objective from the one night attack (El Dorado Canyon) to the extended operation (Desert Shield and Desert Storm). Subsequently, there is a quick return to a peacetime tempo. The spatial aspect of the virtual combat air staff allows staff functions to be performed with equivalent results from buildings (tents to hardened bunkers), vehicles, ships, or aircraft. Both the temporal and spatial aspects of virtuality serve to make it an attractive paradigm for guiding military restructuring plans.

Now that we have reviewed staff history, technological promise, business sector trends, and reengineering issues, it is time to speculate on how all these factors might come together in a realistic scenario. The following account is but one vision for how a virtual combat air staff might find itself operating in the future.

Chapter Six

CASE STUDY: A FUTURE VIRTUAL COMBAT AIR STAFF SCENARIO

Previous chapters in this report surveyed the development of the staff through history, the application of "virtual" concepts in the commercial world, and communications and information technology trends. It is now possible to present a futuristic scenario in which examples of "virtual" approaches could be employed by a combat air staff tasked to conduct an air campaign in the year 2020. A scenario-driven approach is provided in an effort to assist the reader in understanding the synergistic effects of a virtual staff approach to waging air war.

AIR COMPONENT COMMANDER: AIRBORNE OVER THE ATLANTIC

At 25,000 ft over the Scottish coast, the EC-777 banked sharply as it disengaged from the boom after the final refueling. The fuselage of the command and control jet glistened in the noonday sun. The jet accelerated, then settled into its cruising altitude.

Inside the otherwise dark cabin, the soft lights of computer terminals glowed next to reading lights and arrays of advanced electronic gear. Air Force Lt Gen "Cat" Katamaran, the Joint Forces Air Component Commander, slowly lifted his 200-pound frame upright as he adjusted his chair. He checked his watch against the time blinking on the computer screen. He toggled the "stick" to the "flight plan box," mentally noting the blinking countdown time until landing display (CTULD) that told him he would be on the tarmac in less than two hours at Ramstein Air Base, Germany. Cat chuckled as he looked around the cabin, his personal staff feigning activity at their stations as they slowly woke from various states of slumber. The bold among them might grab a few winks if they thought the general was asleep, but everyone was up when the general was awake.

A click of the joystick button retransmitted his e-mail announcement across the world. The video teleconference with his combat air staff would commence in 2.5 hours. The big topic, transmitted to his staff via e-mail yesterday afternoon, would be his proposal to transition from close air support (CAS) operations to the strategic bombing part of the air campaign in the next three days. He made a mental note to check with the separate elements of the staff prior to the meeting to head off any problems early.

Cat scanned the computer screen. Multisource intelligence assessments were stacking up in the "Intel Room," a 3-D virtual workspace located in the bottom right side of the computer screen. In the upper left portion of the screen, the Morning Situation Briefs rolled into the "Operations Room" folder. A "hot" label floated over the "BDA folder," containing unmanned aerial vehicle (UAV) and head-up display (HUD) camera video segments from last night's attacks. Cat toggled the stick down the screen of his workstation JFACC Situational Awareness System (JSAS) Mk IV to the "Campaign" 3-D room to review the campaign history file. He liked to start with the big picture of the air campaign before his staff enmeshed him in the specific details of staff planning.

Cat skimmed through the "EarlyBird" Political/Military Overview, a summary culled from international news sources. World News Network (WNN) carried the best summary, and he loved going to the commercial sources to keep his own intelligence analysts honest. He read the background brief:

> The North African Pan-Islamic League (NAPIL) has finally made good on its threat to invade Egypt. Under the pretext that Egypt's refusal to permit overflights of pilgrims to Mecca constituted an unacceptable restriction on its free expression of religion, the combined units of what seven years previously had been the Algerian and Libyan armies attacked Egypt. Amazingly, the Islamic fundamentalism that swept eastward from Algeria when Mu'ammar Qadhafi's successor, General Abner bin Izzi, embraced the cause in

2012 had been blunted at the Egyptian border by a combination of devastatingly accurate American airpower and a rapidly modernizing Egyptian Army that was now the envy of northern Africa. But Egyptian progress in building a thriving Muslim state was still threatened with complete destruction by "the Scimitar," as the thoroughly disciplined and well-coordinated force from NAPIL called itself. The fundamentalists had also been modernizing— although the emphasis had clearly been on their military forces— and their integrated C^{4}I systems were proving particularly effective.

Cat clicked to the continually updating Daily Situation Report (DSR), and adjusted his headset. He leaned back in his chair as the pleasant voice related the latest developments on the front as a flurry of slides, video, and audio summarized the combat situation and brought him quickly up to speed. Up to this point, day four in the conflict, the air campaign was progressing better than anticipated. Three of the four main armored columns speeding across the Libyan desert had ground to a halt under persistent American air attacks. Only the southernmost armored column continued to limp forward into Egypt. Much of the misfortune of the NAPIL forces was due to the application of American airpower, which seamlessly combined high technology, superior training, and a multitude of information-warfare applications. In particular, the multitargeting precision-guided weapons were proving particularly effective at tank killing.

The big news of the previous day had been a successful NAPIL shootdown of an F-22, although the DSR announcer dispassionately reported this event as though it was not particularly newsworthy. Nothing was further from the truth. The shootdown was at the forefront of Cat's concern: Although it was only the second loss in the war, any casualties were critical. A concerted Marine combat search and rescue (CSAR) effort had retrieved the downed pilot and whisked him back to the *USS Reagan,* cruising in the Mediterranean Sea just north of the African coast.

A blinking alert line at the top of the computer screen notified General Katamaran that his intelligence chief was on the link. Cat moved his microphone closer to his face and mumbled, "Jack," over the EC-777 internal battlestaff net to his communications specialist, his vocal chords still drowsy, although his mind raced.

"Yes, General," Jack replied tersely.

"Jack, I want full data rate for the Intel Link. I want simulcast review of the Lead's camera video with General Spokane. Also, go to maximum encryption, I need more background than I'm seeing in the Intel assessments. Give me 30 mikes[1] to work through the Intel assessments, then put General Spokane through."

"Roger, Sir," Jack replied, pulling up the communications configuration menu on his screen.

Cat sat up, grabbing his coffee cup. He watched the status lights go to green as Captain Jack Reynolds manipulated the configuration panel on his computer to open additional communications channels on the satellite communications (SATCOM) link. Cat leaned far to the left, carefully positioning his coffee cup below the java dispenser. He settled in to read the assessments as he waited for the link to synchronize and General Spokane's face to appear.

The computer quickly supplied him with a representation that allowed him to move directly to a "drawer" where he could call up the intelligence assessments. Sure enough, there were the summaries from the different intelligence agencies, NATO, and his own command intelligence (J-2) briefs, compiled in all of 45 minutes after his initial request. "Not bad," he thought to himself, considering they were assuredly based on information gleaned from covert, national, and military sources collected within the past couple of hours. Still, despite a superior presentation of enemy force structure, placement, and potential maneuver space, he was disappointed in the material on likely motives and plans. While an intricate series of possible follow-on military operations was proposed, the political dimension was sorely lacking. "Seems like no amount of computer power will ever let you get inside the other guy's head," he mused to himself.

INTELLIGENCE CHIEF: ACS/I, DEFENSE INTELLIGENCE AGENCY CENTER, BOLLING AFB, WASHINGTON, D.C.

Maj Gen "Spook" Spokane had seen the e-mail message from General Katamaran directing the change in objectives from CAS to a

---

[1] *Mikes* is military jargon for minutes.

strategic interdiction phase. In response, his targeting cell at Shaw AFB, South Carolina, had completed the database manipulation, worked up the target lists, and finished the target priority recommendations last night. Spook quickly scanned the list, noting that General Cat would probably question the first four priorities and accept the rest of the list without comment. There were few problems transitioning to strategic bombing. Years of reconnaissance pinpointed the major strategic military assets of the NAPIL, and as long as there were no more surprises in the Libyans' surface-to-air missile (SAM) arsenal, the rest of the conflict would be a simple mathematical relationship between the number of targets destroyed and the point at which the NAPIL sued for peace. The campaign planning models predicted ten days to two weeks until resolution. However, there was still that SAM problem from last night's mission.

Spook rubbed his eyes, blinking against the early morning sun as it reflected off the buildings across the Potomac River. He had spent the last two hours electronically connected with his NAPIL SAM cell, whose members resided in various intelligence agencies across Washington, D.C.; at the Joint Command and Control Warfare Center (JC2WC) in San Antonio, Texas; on the Tonopah Range in Nevada; and in weapon labs from Florida to Ohio. Each of the analysts silently watched the F-22 HUD video and the wide-angle UAV overhead movie of the shoot down, trying to piece together their best assessment.

"Sherwood, what do you think?" Spook asked finally, addressing Sherwood Holmes, Chief of the NAPIL SAM countermeasures team at DIA.

"Obviously laser-guided, General," Holmes answered, toggling his computer joystick to produce a pointer on the frozen frame of video. "We took the video apart frame by frame. This is the advanced laser-guided SAM—probably the XLG-7—that we were worried about. Jammer has the countermeasures breakdown."

Maj John "Jammer" Jameson moved forward on his stool at the video conference room at the JC2WC in San Antonio, Texas. He clicked the middle button on his mouse, bringing up a countermeasures slide. "General, as you can see by the timing slide, there was no way to get out of the way of this missile. There was no raw gear indication, no laser warning receiver alarm, no visuals.[2] Likewise, none of the heat pointers except the missile launch return light illuminated. By the time the pilot saw that light, he was directly overhead. There was nothing he could do to get out of the way of that SAM. Because it was a multiple-warhead SAM, it could have taken out the entire flight. The only thing that saved them was a flash of light from the launch that the Lead noticed reflecting off her canopy. That's her voice yelling on the video. She saved them all but number four, since he was in trail. The biggest story is that none of the laser countermeasures worked. I think Dr. Eisenstein has the breakdown on the "trons."[3]

"Doc, what do you have on the electronic analysis summary?" Spook queried, addressing Doctor Allen Eisenstein, renowned MIT physicist on loan to the Lab at Wright-Patterson AFB, Ohio.

"Morning, General," Doctor Eisenstein offered, absent-mindedly scratching his head. "We've been up all night trying to figure out the wavelength of the laser seeker. We knew it was in the visual range, but it's probably a little higher than all our countermeasures. They must have tuned up the laser's wavelength to the upper end instead of the middle of its range. That's the only reason the colors in the color-spectrum analysis don't match the countermeasure tuning range. The spectrum readout matches the XLG-7 readout we saw when we tested a model earlier this year. I think we have a match."

Spook nodded. "I thought the XLG-7 was only in the NATO inventory? Sherwood, give me the XLG-7 background, and I don't have to remind you that our assessment, which had CIA concurrence, on the probability of the NAPIL fielding this SAM in four weeks was low." Spook added, "What the hell happened?"

"You're right, General," Sherwood responded quickly. "Nobody predicted that NAPIL could get it out in the field that fast, but we do have some interesting background information. A dozen XLG-7s disappeared from a NATO warehouse in Germany and appeared at a

---

[2] *Raw gear* is military jargon for the radar warning receiver (RWR) equipment suite onboard aircraft.

[3] *Trons* is military jargon for electrons, a reference to the electronic energy that is detected by the missile attack warning gear onboard aircraft.

covert arms sale in Iraq about two months ago. We tracked the shipment into Lebanon, then Egypt, but lost it. We don't do too good of a surveillance job on our allies."

Spook chuckled. "Yeah, we need to be aware of what our friends are involved in just as much as our enemies these days. Sherwood, walk us through the engagement."

Sherwood clicked to the UAV footage. "General, the UAV footage helps here. As you can see by the two plumes, the NAPIL battery launched two XLG-7s, with the second one missing the flight. As we follow the next few frames, you see the follow-on F-22 flight destroying four more missiles and two launchers. We brought in the fine-grain overhead to verify that the pieces were XLG-7s, and the destroyed missiles' fin paint matches the NATO signature, as you can see from the high-resolution analysis. Further multispectral collection did not locate any other XLG-7s in the vicinity."

"Sherwood, my simple math skills tell me that at most we destroyed six of these XLG-7s," said Spook. "Where are the other six?"

Sherwood brightened a little, wiping a bead of sweat from his forehead. "We played back the footage on a NAPIL resupply convoy yesterday. It allowed us to track the convoy back to Tripoli. We are pretty sure we know where the other six missiles are."

"Good," said Spook. "Don't keep me in suspense."

Sherwood continued, "The bad news is that the missiles are probably stored in the basement of the Arab Youth Recreation Facility." Sherwood pushes a function key, and a color photograph appears. "This is an overhead shot of the building. I also have a HUMINT report that is contributory. Since it is a Category-10 source, I put a copy on the link. Because of the classification, you won't be able to break it right there at the stream, you'll have to pull it out and get the decoder out of the safe and run it through the scanner."

"I'll read the report later," Spook replies. "Are you telling me this building is a soft target?"

"Yes, sir." Sherwood adds: "The day before yesterday, some big trucks were delivering some things, some big crates, that seemed to match those of the XLG-7 missiles. A check of the numbers revealed a date on one of the boxes that coincided with the Iraqi arms sale date. You'll see some other information in the Category-10 report."

Spook breaks in: "What are we going to recommend to the big guy? Can't hit anything nonmilitary with anything hard, and the nonlethal option takes presidential approval. You guys remember when we put that 'Gumball'[4] on that bus station in Brazzaville to quell the rioting in the Congo last year?"

Heads nod across the network.

Spook continues, addressing no one in particular: "You know how much trouble that got us into. Everything located within a few hundred yards around that station was covered in that gunk. Nothing moved for days and the people inside almost starved—a major diplomatic incident. If we drop a Gumball on the Youth Facility, it will keep those missiles out, but someone's going to have to deal with the fallout. Well, I'll brief that to General Katamaran at our staff meeting in one hour, and we'll find out if he wants to push it up the chain or if he wants to keep the fine-grain surveillance and intercept the missiles on the way out. Now that we know where those things are, I think, either way, we'll be able to do something with it. Jammer, give me the VCM [visual countermeasures] options. What's the proposal, software or hardware?"

Jammer pulled up the VCM Overview Slide. "General, obviously we'd like to change out the VCM black boxes as fast as we can, but we are looking at three days. We have the boxes in the warehouse, but the physical shipping parameters are rough. I don't think General Katamaran is going to stop the campaign to wait for the hardware. We think we've got an interim software fix to try to take care of the problem for the next three days, until we get the boxes with the new wavelength parameters to Italy."

"Roger, that," Spook replied. "I'll talk to the Logistics Chief this morning. I'll e-mail him at Ramstein to make sure they have something open for us. When do you think you're going to be ready to ship?"

---

[4]A nonlethal weapon that leaves a sticky bubble gum–like film on everything.

Jammer pulls up a timetable slide. "We'll ship by the end of the day, and if we get the boxes on tomorrow's 0900 flight out of Scott, it will get to Italy by tomorrow night, but too late to help with the combat missions for tonight and tomorrow."

"If that is the best we can do on the hardware side, what about software?" Spook asks.

Jammer pulls up the software change request forms. "Our engineer in Italy pulled the software modules out, so we know what to do to give the F-22s a little better wavelength coverage. It is just an interim fix, but he can make the changes in time for tonight's missions. It will take a good four hours to get all the changes complete, but that's why we're paying all the overtime."

Spook notices an alert line at the top of his computer screen, notifying him of a pending electronic link with General Katamaran. "Okay, Jammer, press ahead with the software changes until we can get the hardware. I've got General Katamaran on the other line. Keep an eye out for any message traffic on the XLG-7 situation. Over and out."

As General Katamaran pulled his coffee cup back, the outline of General Spokane appeared on his computer screen. "Morning Spook. As you can see, I've got you on a full datalink, that way I can watch your lips move so I can tell when you Intel guys are lying."

"Morning Cat," Spook replied, "Glad to see the full datalink, that way I can watch your hands move so I will understand what you old fighter pilots are saying."

Cat laughed, but got right to business. "Spook, I saw the Intel analysis on the shootdown last night. Good thing Lead saw the launch flash. Let's take a look at the video."

The generals reviewed the F-22 and UAV videos. General Spokane outlined the proposal for the software change to buy time until the new VCM boxes got to Italy. General Katamaran agreed with the plan, e-mailed the Logistics Chief, and directed General Spokane to schedule a follow-up test to make sure there would be no capability lost in the lower laser ranges. General Spokane outlined the analysis on the location of the remaining XLG-7s and possible targeting options, including the Gumball, a nonlethal munition. General

Katamaran deferred a decision, but directed increased reconnaissance of the Arab Youth Recreation Facility in Tripoli. The discussion returned to targeting priorities for the strategic bombing campaign.

"Our first deep-strike missions begin tonight," Cat started. "Why are we going right to the processing plants before going to the depots? Doesn't seem to be a logical way to do it if we want to roll up the NAPIL Army."

"The thinking on the targeting staff," Spook explained, "is to kill any chance of any chemical or biological stuff coming in later. If we make Babymilk Plant the top priority, we negate any nuclear, biological, or chemical (NBC) weapon use when the NAPIL gets desperate. Now, it's true that the sortie rate model shows that we add two days of air strikes to the war by hitting the NBC source early, but it might save some real headaches later."

"Go ahead," answered Cat, feigning frustration. "Push that targeting for the processing plants up to Priority One. All we need now is for this to degenerate into a 'gas war' and really make life difficult. Spook, keep working the target list. We want to play it out over a period of at least one week, and see what the modeling run gives us if we adjust to the lower priority targets."

"Okay boss, I'll get the team on it. Thanks."

48TH AIR WING DEPLOYED, SIGONELLA AIR STATION, SICILY

"Jack," General Katamaran called over the internal battlestaff communications net.

"Yes, General."

"Jack, I'm going to the rear to 'adjust gross weight,' so convert me back to regular data rate and standard encryption for my link to General Llewelyn."

"Yes, sir."

Brig Gen Kenneth Llewelyn, Commander of the 48th, had deployed to Sicily three days before, when things looked to be getting bad in a hurry in the North African desert. His deployed air wing included F-15Es, F-22s, F-26s,[5] and a variety of integrated UAV assets.

General Katamaran returned to his seat, slipping his headset back on and clicking the button on the computer "stick." The JSAS screen dissolved into the image of a much-too-young lieutenant, becoming extremely nervous upon seeing the JFACC's face. Cat could see that the room behind the young officer was a lot less active then he would have suspected, but then he remembered this wasn't 1991, he wasn't in the "Black Hole" at Riyadh, and things had changed a lot since his first exposure to a command center as a captain in the Gulf War. That lieutenant commanded more information processing and communications power at his fingertips than the entire Combat Operations Center used in two months in Desert Storm. The "L-T" called out to his boss that he was transferring the general over, and Kenny was quickly on line.

"What's the story, Ken?"

"First look at today's missions is promising, sir," replied the gruff and overworked forward commander. "The good news is that we're starting to tear up that southern column of armor. Our advanced mission planning system is down. We can't tell if it's a penetration, but we found some spurious characters in a mission simulation program that deleted some known NAPIL AAA sites. We're lucky the system software is smart enough to let us know there are anomalies and that we're not pushing ahead on bad dope. Still, my OPSEC [operational security] guys are going to have a lot of explaining to do. We should be back up within the hour with no loss of data, just in time to see what sort of damage the 'Bones' [B-1s] from Dyess inflicted on that southern NAPIL force. Thank God that we can still get them targeting data with our backup links."

"How much more do you think you can do today?"

"Well, we'll have executed today's ATO plus all its real-time revisions by 1400 Zulu. Tonight's attacks should be plotted and distributed by midafternoon if the system's up and running again soon, but we can always ask the *Reagan* to distribute the breakdown if the mission

---

[5]Fictional designation for the Joint Strike Fighter.

planning system doesn't come back up. I'm mostly concerned about getting good BDA so the system can generate something of optimal use to us, but those bastards have already taken out some of our on-scene 'eyes.' Looks like they got some sort of multiwavelength laser that temporarily blinds the optics on our old Predators. We're going strictly with SAR [synthetic aperture radar] imagery and the high-altitude stuff now. It's slower, and the fidelity isn't quite the same, but we're making do. I don't think our tempo will suffer at all."

"Good." Cat replied, "Tell me about your folks on Crete."

General Llewelyn's face reddened. "What happened on the Intel assessment for this XLG-7 yesterday? Did you see the video? Man, that thing could have taken out the whole flight with its multiple warhead. They were pulling some pretty good Gs. If Lead hadn't seen the reflection off the canopy top, those bastards would have splashed four instead of one. . . ."

Cat expected the tirade, but let Ken vent his anger. Cat reminded Ken that Intel had recommended only night operations until NAPIL enemy air defenses were significantly attrited, but the need to blunt the NAPIL offensive forced more-aggressive action.

Ken, a little more settled, continued: "As you saw in the SITREP [situation report], one other jet took some damage from the shoot-down and declared an IFE [in-flight emergency], so we diverted the flight into that little NATO base in Crete. Our maintenance guys are there and should have the jets ready to fly back by tonight. We're watching the political indicators so we won't fly out until cover of darkness. I talked to the Lead last night on a secure link from her Palm-max computer. She was upset about the shootdown and was trying to blame herself. . . ."

"I signed the DFC [Distinguished Flying Cross] on her," Cat broke in. "I also had Personnel flag her records for the PACAF operations officer selection board this fall. With a DG [Distinguished Graduate] from Weapons School, Number 2 at William Tell, and this DFC, I have a feeling she is going to see "major's pay" a couple of years earlier than the rest of her year group. If you feel like you want to keep your folks in Crete for another day, don't sweat it. Let them get a little R&R on those Greek beaches if you can spare the manpower. What is the schedule on Powers' surgery?"

"The Marine chopper was in quick and picked up all the important pieces," Ken answered. "Powers had a small burn on his neck and caught a piece of metal in his side, so he's scheduled for surgery any minute now. My flight surgeon and the Navy surgeon are cutting him open on the *Reagan* with a telemedicine feed from the specialists at Walter Reed [Medical Center in Washington, D.C.]. In fact, if you don't mind, I was going to chopper in to the *Reagan* this afternoon to be there when he wakes up."

"Good idea," Cat replied. "Make sure you have the advanced Palm-max in case I need you. And if the kid is up to it, get a video link back to his wife so he can talk to her, and she can see he is all right. I'll talk with Logs[6] at Ramstein when we land in about 30 minutes and get the datalink patched. Since we've got a teleconference with everyone in about 45 mikes, I'll let you get some work done before we start the meeting. I especially want to know if you need anything special as we transition to the strategic bombing portion of the campaign. I've seen enough tank-plinking video, and besides, its time we let the Egyptian Army mop those guys up so they can take all the credit. Out here."

General Katamaran unplugged his headset and reached for his Nomex gloves. He stood up and stretched. "Time to show this young IP [Instructor Pilot] in the right seat how to land this fully loaded 'triple-7,'" he thought, slipping between the battlestaff seats on his way to the EC-777 flight deck.

MANPOWER SUPPORT COORDINATOR: PERSONNEL AND FINANCE INTEGRATED SUPPORT CENTER, RANDOLPH AFB, TEXAS

Col Ken "Staff" Stafford monitored the steady stream of e-mail from deployed locations. The traffic had been especially heavy from over the Atlantic Ocean that morning. He checked his list. The medal request from Italy had come in, been formatted and forwarded, electronically signed by General Katamaran, approved at the JCS and, with the citation, forwarded back to Italy. That F-22 pilot was going to receive a DFC when she returned to Italy that night. The coordi-

---

[6]Nickname for the Logistics Chief, Brig Gen Ron "Logs" Loggersen.

nation time for the award was four hours, not bad considering that decorations once took months to process. Staff also saw the flag on her records and added the database pointer for the PACAF operations officer board.

Colonel Stafford was most concerned about the financial projections. He had a pending contract for a forward-based contractor F-26 depot facility at Sigonella, and the Navy was having a hard time coming up with its portion of the funds. He composed an e-mail for General Katamaran to coincide with the end of the staff meeting to get some bigger guns in the fight. He composed charts with the financial totals for all aspects of the operation. The financial projections were on schedule. As he modeled the projections he noticed "cross-over times" at the four week point. If the campaign went longer, a congressional stipend would be needed to make up the differences. It was just too close to the end of the fiscal year to find new money. This depot facility contract wasn't going to help the money situation. Stafford made a mental note to talk to the Pentagon, maybe they could free up something like a contingency account.

Just then the e-mail came up describing in detail that the operations would be transitioning to strategic bombing. Stafford called up the manpower documents. A change of billet structure would be required, due to the utilization of advanced precision-guided munitions (PGMs), which required special handling, instead of cluster munitions. Stafford called up the deployment and mobility documents. In Italy, the transportation unit had been there for three months and was due to rotate out in another week. He pulled up the World Wide Notice of Requirements and put a notice of those jobs out in the world to be filled. The supply guys wouldn't be too much of a problem, they were already set and ready to go. Just needed to match the names with the billets. It would be a little bit different with the transportation rotation though; virtually everyone was at the maximum for authorized TDY [temporary duty] days for the fiscal year already, and they still had four months to go. Stafford figured he would start working rotations to see who would be coming out of training school in the next few weeks. There was a new class of transportation specialists due to graduate. Maybe he could convince a transportation unit commander to send a couple of green guys over with the replacements. The e-mail barrage continued.

LOGISTICS BOSS: EUROPEAN LOGISTICS CENTER, RAMSTEIN AIR BASE, GERMANY

Brig Gen Ron "Logs" Loggersen nervously crossed his right leg over his left as he propped his notebook computer on his lap in the cramped cab of the green pickup. After a few minutes of starring down the Ramstein runway, he shifted his focus back to the multi-color screen. General Katamaran would be landing in a few minutes. General Loggersen recognized "the Cat's" distinctive voice on the Ramstein airfield approach frequency on the radio located to his left. Logs hated these VIP visits, preferring to work the logistics issues for the latest military campaign electronically. There was enough to do at Ramstein without worrying about the "red carpet" treatment. At least this visit would be an operations stop only, with the video conference and tour of the weapon assembly facility and not a lot of protocol. The Ramstein Base Commander in the staff car parked directly in front of his pickup would handle the lunch and welcome.

Logs called up his top weapon coordinator, Col Bill "Bangs" Gunn. The computer beeped and flashed as it established a secure link across the base, Colonel Gunn's image appeared on the screen.

"Bangs, the general is about ten minutes out," Logs relayed, "Tell those kids behind you to stop moping around, and pretend they know how to assemble some hard-kill PGMs."

"Yes, sir," Bangs laughed. "You know Chief (Master Sergeant) Jones got his reputation as Mr. Clean from preparing for these short-notice visits."

"Tell the Chief to find us one of those damn 'Gumball' NLWs [nonlethal weapons]," Logs replied. "I still haven't figured out why Spook sent us that request earlier today. We have our hands full building 'bunker busters' without worrying about NLWs. Any word back?"

"We confirmed that there are no 'Gumballs' in Europe about 15 minutes ago. I even called each site personally to verify their database listings," Bangs answered. "All the 'Gumballs' were shipped back to the States last year after that fiasco in Africa. Colonel 'Staff' sent me another e-mail this morning on the financial picture, and it looks grim if we really want to pay to get one to Italy. If you want my opinion, boss, this campaign transition couldn't have happened at a better time: Our stocks on the anti-armor clusters are almost down to 'red-line.' Our folks are already hard at work replacing the advanced tank-killing munitions that were already in the pipeline and starting to push the advanced precision-guided depot and hard-kill munitions."

"Well, keep them working the weapons, and keep their hands off the mops. I've got to check with the supply coordinator at Scott on some black boxes, so let me go."

Logs pulled up the supply icon, as Bangs' image disappeared from his computer, and patched into the supply wide-area network. He saw the alert message request for shipping ECM boxes. Another hard problem, he mused to himself; the manifests for dedicated airlift from CONUS were already full, so he needed to make sure he was pulling off some lower-priority things. Those transportation guys in Sigonella were going to have to go another two weeks without tires for their K-Loaders. If you pushed the tires off, it wouldn't help them if they got a flat. Of course, if those ECM boxes didn't get into the jets, it would be hard to explain why they had plenty of spare tires and were losing airplanes.

Logs had an idea. He pulled down the ECM e-mail and toggled down to the shipping request. The file included a graphic representation of the dimensions of the ECM boxes: 48 boxes, each 1 ft^2. Just as he thought. He would ship the boxes commercial. With two button clicks, he pulled up the commercial proposal from the Air Force contracted shipper, Federal Overseas Parcel. The price quote was a little high, but the guaranteed shipping time was four hours better than via Air Force airlift. He forwarded an e-mail to General Llewelyn at Sigonella. With Ken's support, which was guaranteed after last night's shootdown ("Have you seen the film?" "Of course we've seen the film." "What are we going to do about it? We can't exactly stand down even though it seems like three of the columns have come to a stop."), the finance guys would be scrambling for once to find the money rather than his guys struggling to get the job done on the cheap. Speaking of Finance, he had better take another look at that depot facility contract.

He pulled up the proposed contract with the Italian support personnel for moving the depot facility from tents into a permanent hangar. He figured the Italians were looking to renovate the hangar anyway, but he didn't see any way around it. This thing has been promised. He tentatively put his okay on it, electronically transmitted it back up to Randolph and the Pentagon for those guys to hammer out the money problem.

The roar of jet engines jolted Logs back into the present. The pickup was moving down the taxiway to intercept the EC-777. Logs took one last look at General Cat's itinerary: a staff meeting from the VTC room at Ramstein and a tour of the weapon assembly. Show time for takeoff was in two hours. He signed off and disconnected the power and transmission cords from their slots in the dashboard. He slid the computer into the briefcase. The pickup slowed to a stop next to the sleek EC-777. Logs stepped toward the jet as the flight-deck ladder descended. He watched General Cat slip down the ladder as his entourage followed. Seeing the general in person made him realize that it had been a long time since their last meeting, despite their many electronically supported interactions coordinating efforts in a number of past air operations.

"Logs," General Cat exclaimed, picking the one-star from the line of meeters and greeters. "Your blond hair has gone gray, and I never would have known from the other side of the datalink."

Both men laughed.

"Logs, now tell me all about 'Gumballs,' I've got some ideas . . . ."

SOME OBSERVATIONS

The case study highlights some interesting issues germane to the discussion of a virtual combat air staff. The underlying need for vituality remains the desire of warfighters to reduce the number of people placed in harm's way. Within the case study are multiple references to the use of advanced technology components that permitted the use of distributed, and virtual, interactions to accomplish the mission. Such things as VTC, Palm-max computers, and multiple channels of information presented in an intelligent manner are rooted in computer and communication technologies coming to fruition today. One can only predict that additional advances will be made and that additional tools, more robust and intelligent, may be available to implement a more streamlined military force response in the future. In some respects, the application of these tools may reduce the exposure of personnel in those areas that are usually most heavily populated—the fixed bases and planning centers for an air campaign.

Illustrating the working relationships among the characters in the case study, Figure 6.1 provides a simple staff-interaction diagram that captures the interactions among the principal characters of the case study. One might conclude that, below the level of the JFACC, the organizational relationships resemble a network rather than a hierarchy.

Similar to the JFACC sharing of command discussions, a complex form of interaction can also be depicted for the intelligence flow. Such an arrangement is shown in Figure 6.2.

In both examples, the retention of a centralized command authority is evident. But the actual exchange of information and discussions revolving around the issues at hand approaches the network organizational form. With such a tension at work, the need for a hybrid form within the virtual combat air staff concept is evident.

In these examples, we simply show how functions were performed, people worked with each other, and the virtual combat air staff accomplised its tasks. They were able to do so even while transiting to an evolving contingency, addressing the full range of operational, intelligence, logistical, and personnel issues in the process. This is the promise that a virtual combat air staff holds for the future through the application of evolving computer and communication technologies.

Some issues remain, however. As we are beginning to understand, the application of advanced computer and communication technologies also carries with it the potential for additional vulnerability to our adversaries. For that reason, careful consideration of what technologies we adapt and how we apply them to our operational needs is necessary. The case study did not explicitly point out the vulnerabilities of the communication links that may exist, but those

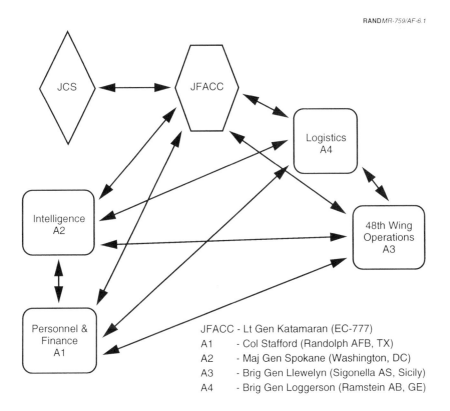

RAND*MR-759/AF-6.1*

Figure 6.1—Virtual Combat Air Staff Communications Net

issues are at the heart of ongoing information-warfare discussions today. They will not abate in the future.

In another realm of thought, the issue of exchanging information with allied and coalition partners will not be solved through the mere application of technology—policy will also need to be changed or developed. Similarly, we should not look on technology as the answer to divining the intentions of enemy commanders. Many of the fundamental things we do today to facilitate successful military operations, such as trying to anticipate enemy actions, will remain in the information age. But the considered application of these

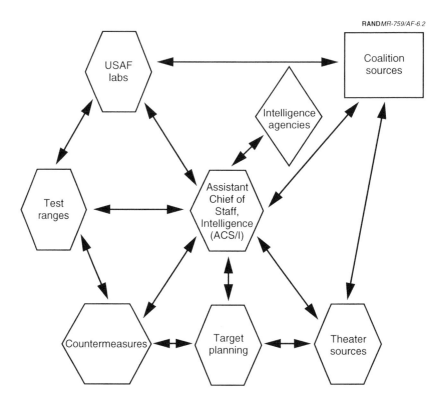

RAND*MR-759/AF-6.2*

Figure 6.2—Intelligence Coordination Net

technologies may facilitate our accomplishment of those tasks and enable us to exert the presence and power of a larger, virtual force into a future of reduced physical assets.

The point to make here is that the future of the military is one of reduced manpower and equipment, and a more automated and linked community that relies on advancing technologies to maintain a credible fighting force. As such, the case study shows one postulated example of how that might be accomplished through the use of a virtual combat air staff to support the air campaign portion of the overall effort.

Chapter Seven

ANALYSIS AND RETROSPECT

Recall that the objective of this study was to look at a number of questions and to explore the ramifications of possible answers: How will military staffs evolve? What adaptations of technology to warfighting operational needs are realistic? How can we adapt technology to achieve a virtual staff? How should the Air Force be structured to achieve a flexible force?

In terms of the military staff function, planning as a function of combat operations is, and will remain, an essential part of the overall warfighting effort. It is not debated whether a planning staff will be required but, rather, how that staff can be formed to take advantage of technological advances that will permit the reduction of the staff component in the theater of operations. Reducing the number of personnel subject to hostile action is a continuing goal of military leaders. Using a virtual concept of combat air staffing may be only one means to achieving that reduction in physical presence while maintaining the ability to plan and support the large and complex task of waging an air campaign.

External pressures will continue to militate for new and innovative ways of accomplishing the military mission—in all of its forms and manners of effort. The reduction of U.S. bases overseas has eliminated the robust basing structure previously used as a jumping-off point for combat operations around the world. Concurrently, personnel and combat planning assets have also been reduced, affecting the ability of a forward-deployed commander to rely on a robust theater CINC staff for support and assistance. This trend will likely continue as U.S. forces are increasingly positioned in CONUS.

Reduced manpower levels are also affecting the resources available for combat air staffs. While the USAF has achieved its assigned downsized level, there are rumors that further reductions in overall force size are coming by the turn of the century. As a result, fewer personnel will be available to continue pursuing the missions of the Air Force around the world. Leveraging those precious resources may only be possible through the application of virtual networking and the rapid exchange of information to enable individuals to benefit from the collective knowledge of the whole Air Force.

Finally, budget reductions, which underlie trends toward a CONUS-based force of reduced numerical strength, may ultimately dictate that the Air Force find ways of accomplishing its missions without resort to use of resource-intensive options. Rather, it might have to transition to reliance on an information-based force, applying its limited weapon systems with high levels of accuracy. Similar to today's concept of multiple target kills per sortie (in contrast to yesterday's multiple sorties per target kill), a single virtual combat air staffer, supported by a host of information sources, could achieve the results that today's larger air operations center staffs achieve. This is the promise held up by adapting future technologies to combat operations—maintaining today's might and presence with reduced physical plant and resources.

What are the advantages inherent in such a promising future? If the virtual combat air staff concept could be incorporated, a reduction in the forward-deployed footprint of forces in theater could be achieved while retaining the ability to plan fully and to execute an air campaign. This reduction would have several effects. First, a larger proportion of U.S. deployed forces would be dedicated to inflicting damage on the enemy, rather than to providing support to combat forces. Second, fewer friendly forces would be exposed to hostile fire. Third, the reduced size of the staff would enable more direct interaction with the key functionaries of the air campaign process, probably reducing some of the "fog of war" and bureaucratic impediments that get in the way of efficient combat planning and execution. Obviously, these advantages will require training and exercises to evaluate whether they are indeed achievable as proposed here.

The future of the information age promises unlimited access to means for communication and the ability to pass large amounts of

information among those who need data. Even without using the
virtual combat air staff concept, the shift in combat operations to
transferring and using large amounts of data will drive the existing
combat staff to use network forms of interaction to achieve the goals
of the air campaign process. During Operations Desert Shield and
Desert Storm, "workarounds" were a daily occurrence that were used
to get the job done. We suggest, by way of the case study, that elimi-
nating restrictions to combat air staff interactions will naturally lead
to network operational schemes that could continue to exist within a
command and control hierarchy to guide the overall process. In
effect, a hybrid form of interaction will be created to accomplish the
mission.

However, the parallel dependence on multiple points of information
exchange creates a vulnerability that must be considered when
approaching the development of a formalized concept for virtual
combat air staff operations. Such a dependence enhances the
potential of hostile information attack, as well presenting the possibility that parts of the military force may now have information tools
at their disposal that may allow them to act in ways previously not
considered. To both protect against information attacks and preclude rogue application of information-warfare tools, clear rules of
engagement must be established that consider cyberspace, as well as
the other traditional dimensions of warfare.

Even without external pressures, a network is a good way to organize
to facilitate information transfer. Because of the dependence of
combat operations on multiple individuals and organizations at different locations, the robustness that results from multiple points of
information residing in various locations guards against the unin-
tentional loss of information. However, multiple paths of communication, attack detection tools, education about potential problems,
and a fighting force that has grown up with computers will help
ameliorate those affects.

Information is the key to the future. Witness the multiple, emerging
operational concepts that rest on a foundation of dominant battlespace awareness and knowledge. Acknowledging that some sort of
hybrid organizational structure best supports combat operations
begins the process of evaluating how a more formalized doctrinal

approach to a virtual combat air staff can be developed. This study is but a first step in that direction.

OVERVIEW OF TECHNOLOGY TRENDS

The U.S. military is developing a vision for the future, a vision that emphasizes optimum command, control, and communications capabilities that will evolve into an integrated global network called an "infosphere." The infosphere will provide worldwide communications support to joint forces through a web of military and commercial systems that link and provide easy, but secure, access to information databases and fusion centers (Meadows, 1995, p. 25). The creation of the infosphere, with its heavy reliance on automation, will allow military units to maximize their use of what is perhaps the asset in shortest supply: time. To the extent that such organizations can get access to data wherever it resides, translate it into meaningful information, disseminate it quickly, and display it clearly to those that can exploit it, the promise of information technology will be realized (Krepinevich, 1994).

To operate in the infosphere, one can reasonably imagine advanced workstations having both audio and visual capabilities and screen-sharing protocols to allow individuals to work together in groups as intimately as if in the same room. Individuals and organizations will have access to information sources located anywhere in the world, along with the means to help them locate and access such resources. Data streams produced by instruments and sensors around the globe will flow into distributed databases and be automatically processed into usable form. Wireless communications will enable mobile forces to maintain their links while on the move (Denning, 1989, pp. 432–434). Multilevel authentication protocols will provide secure communications and, failing that, issue notification of signs of compromise.

Systems of this ilk will become possible through advances in a wide variety of technologies. Opto-electronics, common user interfaces, multimedia front-ends and interfaces, very-large dynamic database systems, natural-language processing and understanding, adaptive learning, image exploitation, intelligent fusion, robotics, machine vision, and many more technologies will be employed to create a ubiquitous knowledge environment (Robinson, 1994, p. 25).

Several years ago, IBM estimated that by, the mid-1990s, one billion computers around the world would be connected by computer networks (Bankes, 1992, p. 10). If one uses successive generations of the most powerful commercial machines as a benchmark, computing cost has been halved approximately every three years (Tesler, 1991, pp. 87–88). Such ever-widening presence of information technologies coupled with decreasing costs has led many to believe that we are fundamentally altering the nature of human transactions throughout most of the developed world (Bankes, 1992, pp. 3, 10). A "relentless compounding of capabilities has transformed a faint promise of synergy into an immense and real potential" (Dertouzos, 1991, p. 63).

To provide a better understanding of those technologies and the technical background to this study, the following will present current information on microprocessors, bandwidth, throughput, advanced processing algorithms, new displays, changing data-presentation formats, mobility of systems, security applications, and the impact of the commercial world on these developments.

THE MICROPROCESSOR

Without doubt, the element most responsible making the exponential growth of the infosphere possible is the microprocessor:

> In the past decade alone, measurement of the information revolution in almost any dimension—*number* (of telephone circuits, television receivers, video recorders, video cameras, or facsimile machines), *capacities* (of transmission media, storage devices, or displays), *speed*, or *cost*—is described not in mere percentages, but in factors of three, ten, or more. (Bankes, 1992, p. 5.)

The 32-bit (and soon 64-bit) version of the microprocessor is making network computing the paradigm of the 1990s just as the transistor enabled mainframes in the 1960s, simple integrated circuits (ICs) provided the technology required for the minicomputer in the 1970s; and the first microprocessors led to the proliferation of personal computers in the 1980s (Burger and Holton, 1992, p. 137; Geppert, 1995, p. 35). The microprocessor benchmark will soon be—if it is not already—the 200 or 300 MHz, 2-million-transistor chip made with submicrometer design rules. By the turn of the century, the bench-mark may very well be tens of millions of transistors on a chip implemented with technology below a quarter micrometer, provid-ing clock speeds in excess of 500 MHz. Access times could shrink from nanoseconds to picoseconds. Capability on the order of 1 billion instructions per second (BIPS) may become a reality. Memory-chip technology will see the 256 megabyte dynamic random access memory at the turn of the century and eventually 1-gigabyte chips as the commodity paradigm. (Shur, 1993, pp. 103–104; Burger and Holton, 1992, pp. 137–142; and Geppert, 1995, pp. 37, 39.) See Figure A.1. By the year 2000, it should be possible to store a VHS quality, one-hour movie on five 1-gigabyte chips—just 250,000 chips could store every Western movie ever made (Negroponte, 1991, p. 109). Some analysts claim that "conservative extrapolation" of current trends suggests that "removable hard disks (or nonvolatile memory chips) the size of a matchbook" and storing about 60 megabytes each will become possible. "Larger disks containing several gigabytes of information will be standard, and terabyte storage—roughly the data content of the Library of Congress—will be common." (Weiser, 1991, p. 101.)

The result is that, by the end of the decade, the distinction between personal computers and workstations will cease to be a meaningful one. Machines will simply be nodes in networks with computing power and data sources available well beyond those residing at any individual node. Architectural innovations that require high com-plexity (e.g., parallel processing and neural networks) will become available to support higher performance at the system level. (Burger, 1992, pp. 138, 142.)

The manner in which this phenomenal progress will be maintained is not clear. As some analysts see it, there is no reason why the revo-lution in microchip capability cannot continue using new materials (such as gallium arsenide, silicon carbide, and even diamond) and

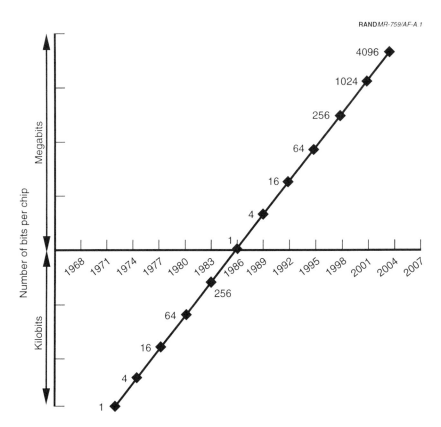

SOURCE: Bernard H. Boar, *The Art of Strategic Planning for Information Technology*, New York: John Wiley & Sons, Inc., 1993, p. 65. ©1993 by AT&T. Reprinted by permission of John Wiley & Sons, Inc.

Figure A.1—Memory Chip Density Timeline

new fabrication techniques. In fact, one analyst expects the past doubling of computer capacity every two years to continue for the next few decades. Such a trend would yield about 20 doublings over the next 40 years. (Canavan, 1993, p. 50 and Shur, 1993, p. 104.)

Some experts see chip and process technology beginning to push against physical limits. Such limits include the size of the atom, the inherent variabilities in semiconductor crystal lattices, and the decreasing levels of transport charge in microchip structures. In the view of such people, "the limits of traditional trends in technological advancement have been reached in the fourth decade of the IC." Nevertheless, optimism is little diminished as experts see new ideas coming to the fore. Currently, packaging is the driver for performance growth (Burger and Holton, 1992, pp. 138, 146, 147). Other, advanced technologies, such as superconductivity and electro-optics, are being evaluated for complex high-performance next-decade chip applications. Opto-electronics is already a reality at the system level (in the form of optical data busses), and progress is being made toward using light, rather than electrons, to transmit information between chips (Barron, 1992, pp. 144–145; Bankes, 1992, p. 9).

In the past, the main determinant of processor performance has been density: the number of transistors that can be put on a single chip. Plotting either processor density (number of transistors) or performance as measured in millions of instructions per second (MIPS) through time (Figure A.2) has allowed analysts to predict the future for microprocessors giving a relationship known as Moore's Law. Essentially, this law predicts an order of magnitude increase in density or MIPS every seven years. Emphasis on new architectural design rules, including the reduced instruction-set computer and parallel instruction execution, is allowing progress to be made to break slavish obedience to this Law. ("Moore's Law Meets MIPS," 1992, p. 125.)

Another emerging technology—which is very early in its development but holds promise to accelerate microprocessor performance greatly—is nanoelectronics. Nanotechnology involves computing on a near-atomic scale and takes advantage of quantum effects to create new types of devices. The multistate stability of quantum devices offers the possibility of multivalued logic functions at greatly enhanced speeds. If the promise inherent in this technology pans out (in perhaps 20 or 30 years), computers based on these principles "would make even today's best machines look clumsy and plodding by comparison." Such machines would offer "orders of magnitude improvement in speed and density over conventional logic schemes" and, through the merging of logic and memory functions into self-sufficient circuits, would provide a new form of intelligent memory. (Wind and Smith, 1992, pp. 140–141; Roos, 1994, pp. 31–36; Shur, 1993, p. 104.)

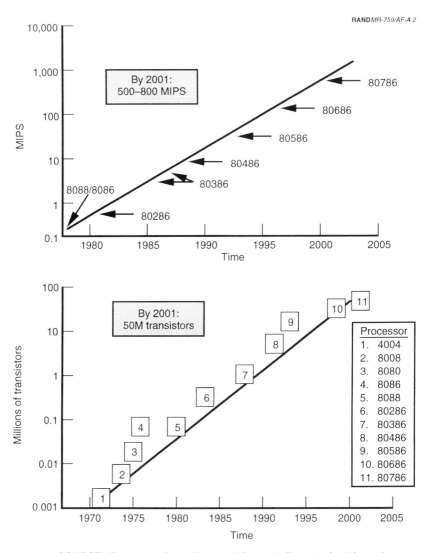

SOURCE: Bernard H. Boar, *The Art of Strategic Planning for Information Technology*, New York: John Wiley & Sons, Inc., 1993, p. 55. ©1993 by AT&T. Reprinted by permission of John Wiley & Sons, Inc.

Figure A.2—Intel™ Microprocessor Family MIPS and Chip Density Forecast

DATA RATES AND THROUGHPUT

A key concern in the implementation of any system(s) proposed to support the virtual staff is the requirement for high date rates. Throughput rates will be very high to support such services as video teleconferencing, database access, rapid data transfer of very large files, and imagery transfer and manipulation, for a large number of users in-theater, many of whom will desire to be mobile. While it is difficult to determine even ballpark figures for what the total bandwidth requirements will be, a look at commercial requirements and developments in the supporting technology can provide some feel in this regard.

The need for broadband speeds can be traced to a number of sources. See Figure A.3. At 10 Mbps—a respectable throughput by today's standards—moving complete files of digitized newspapers, magazines, books, and other textual information over a network requires inordinate amounts of time and expense. Broadband access to file servers is required to permit rapid scanning of visual and textual data. The new field of teleradiology, in which one transmits medical imagery, requires rather high throughput to transfer enormous files. (Kleinrock, 1991, p. 114.)

Cellular telephony is now conducted at 9 to 14 kilobits per second (Kbps), with other voice and data services provided at about 64 Kbps. Current portable wireless communications using infrared transmissions can achieve one Mbps, while radio devices can operate at 2 Mbps. The Ethernet provides 10 Mbps, fast Ethernet 100 Mbps, and the emerging asynchronous transfer mode (ATM) over 150 Mbps (Forman and Zahorjan, 1994, p. 40). Such upper-end speeds are important because, at 100 Mbps and with the use of compression technology, networks are able to handle movies, TV, game graphics, and text at acceptable speeds (Lewis, 1994, p. 63). Nonetheless, the near future will see requirements for 150 to 620 Mbps to support integrated voice, data, and image services (Kleinrock, 1991, p. 114).

Without a doubt, optical fiber will be the medium providing these large bandwidths. Millions of miles of glass already handle most long-haul communications and are capable of relaying data at speeds of over 2 Gbps (Dertouzos, 1991, p. 63). At such speeds, a

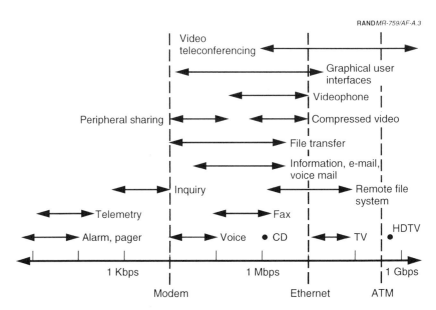

Figure A.3—Application Data Rate Requirements (bps)

one-hour video can be shipped in only five seconds if use is made of digital compression techniques (Negroponte, 1991, p. 109). By mid-1998, upgrades to "optical transmitters and receivers could quadruple [this] data rate." (Brody, 1995, p. 28.) In this vein, telecommunications engineers and scientists have been working to perfect "wavelength division multiplexing", which, according to Herb Brody, Senior Editor for *Technology Review*, is a technology that will allow a single optic fiber to carry multiple channels of information via laser (Brody, 1995, p. 28). Systems based on this technology should be available by the late 1990s and could have a capacity of 40 Gbps, according to Vinton Cerf, a senior vice president at MCI's data services division (Brody, 1995, p. 28). Coupled with these outstanding bandwidth capabilities is a decided cost advantage. For example, it is now less expensive to install fiber than it is to install copper for large office buildings (Kleinrock, 1991, p. 114).

According to VCJCS Admiral Owens, by 1998, commercial firms "will have strung fiber-optic cables to connect 98 percent of the world's cities with populations of 100,000." (Matthews, 1995, p. 24.) In fact, Navy sources relate that: "Undersea fiber-optic voice channels have mushroomed in number from about 100 million in 1980 to about 10 billion today, and should climb to nearly 100 billion by 2010." (Holzer, 1995, p. 3.) Obviously, we are in the midst of the creation of a fiber-optic network that extends well beyond U.S. borders. The military potential for such a system is clear, and as Admiral Owens has said: "We may want to stay plugged into the fiber-optic network when we are doing war-fighting because it can send an awful lot of data and information to us."[1]

A major issue facing the establishment of a worldwide system is "not the lack, but the multiplicity, of standard interfaces between the hardware, software, and network platforms." It is still too early to tell what the eventual universal standards might be, and "the plug-and-play era of client/server systems is many years down the road." (Lewis, 1994, p. 59.)   According to Brad Friedlander, Senior Consultant with Arthur D. Little, "it will be five to 10 years before we get any universal high-level protocols. Sooner than that, companies will develop their own proprietary protocols."(Nance, 1991, p. 272.)

In the meantime, multiple servers will be configured to coordinate with each other so that, in effect, the network becomes "one large disk drive." Furthermore,

> transparent data access across mainframes, minicomputers, and personal computers [will] become a reality. . . . You'll have better interfaces and underlying expert systems that monitor, diagnose, and care for your network. (Nance, 1991, pp. 268–269.)

Evidence of this progress can be seen in the design of the Broadband Integrated Services Digital Network (BISDN) system. Employing ATM switches and synchronous optical network (SONET) fiber transmission technologies, the BISDN could very well become "the 21st-century equivalent of the 20th-century telephone network." (Cerf, 1991, p. 78.) The BISDN holds the potential to create a com-

---

[1]Admiral Owens, as quoted in "Exploit Civilian Technologies Says No. 2 US Military Leader," 1995, p. 19.

mon network for all information and communication services, obviating nets dedicated to such different services as voice, data, and video. ATM helps the BISDN achieve its objective by permitting data packet switching at extremely high rates—the kinds of rates needed to process the enormous information transfers envisioned for tomorrow's users. SONET standards for a hierarchy of optical transmission speeds in excess of 13 Gbps help meet BISDN bandwidth requirements (Kleinrock, 1991, p. 114).

Other technologies that should have significant impact are erbium-doped fiber amplifiers and wavelength division mutliplexing. These technologies could very well transform lightwave communications. Such devices are broadband-capable possessing more than 1 terahertz of bandwidth. AT&T has already demonstrated in the lab that such capabilities are realizable, having transmitted data through optical fibers at a third of a terabit per second. (Kobb, 1995, p. 30.)

USAF laboratories are working hard in this area too. Rome Laboratory, at the Rome Air Development Center, New York, has set its sights on improving signal processing system throughput by a factor of 100 every seven years. Enhancement by a factor of 10:1 in throughput performance in distributed computing and achievement of real-time performance in multimedia databases are priority goals. (USAF Rome Laboratory, 1994.) Wright Laboratory, Wright-Patterson AFB, Ohio, hopes to improve fielded processor throughput 20 times without affecting fielded software, thus saving $1.5 billion in costs. (USAF Wright Laboratory, 1994.) Phillips Laboratory, Kirtland AFB, New Mexico, is working to "leverage high-efficiency, lightweight signal processing technology to reduce the weight, volume and power requirements of space communication systems by more than half." Such improvements would make "direct broadcast of battlefield information to tactical forces feasible and . . . provide affordable C^3 capability for worldwide theater operations." (USAF Phillips Laboratory, 1994.)

Military satellites are beginning to demonstrate the benefits of this research. Newer satellites are achieving the significant increases in bandwidth that accompany operations at higher frequencies. The superhigh frequencies (3–30 GHz) are in use by the DoD's DSCS III and UHF Follow-On (UFO) satellite constellations, as well as leased circuits on commercial satellites at C-band and K_u-band. However, in the EHF realm (30–300 GHz), bandwidths begin to approach the needed levels, and MILSTAR and other follow-on systems will operate in this domain. The MILSTAR system uses 44 GHz for the uplink, 60 GHz for the crosslinks, and 20 GHz for the downlink. These frequencies permit megabits-per-second–class throughputs. (Jain, 1990, pp. 1177, 1182, and 1186.) In general, while emerging communication satellites will continue to show remarkable progress, they will continue to lag terrestrial systems by orders of magnitude in terms of raw throughput. A significant drawback is that anti-jam capabilities can often be achieved only at some sacrifice to throughput capacity.

Dr. David T. Signori, Jr., associate director of DISA, has stated that surge requirements for wartime communications are a key concern, especially for satellites (Ackerman, 1994b, p. 70). According to an article in *Aviation Week*, DoD analyses see a shortfall in addressing user demand for high–data-rate connectivity. In the 2004 time frame, the warfighter will require data rates greater than 4 Gbps via satellite communication links. As Maj Gen William J. Donahue, director of command-control systems for U.S. Space Command, related, there will be

> a huge information appetite in future weapon systems and support functions. . . . Just-in-time logistics, intelligence, command and control—all these are driving information requirements up bigtime. But we'd rather move information than move support troops to the field. (Scott, 1996, p. 62.)

A possible solution to this problem would be to leverage new commercial systems that are due to come on line around the turn of the century. The so-called "Little LEO" (low earth orbit) systems will transmit using frequencies below 1 GHz to provide electronic mail and paging to portable and mobile devices. The most ambitious Little LEO system has been proposed by Teledesic Corporation, whose satellites would relay data packets to one another at rates as high as 1.244 Gbps. Satellite voice telephone service will be provided by the Big LEOs—large, elaborate mobile satellite systems—operating above 1 GHz. (Lowe, 1995, p. 29.) The geostationary Spaceway system, from GM Hughes Electronics, promises to offer all-digital voice, data, and video services within a couple of years. "Subscribers would receive telecommuting, telemedicine, digital libraries, and other high-bandwidth services at data rates of 16 Kbps to 1.544 Mbps." (Kobb, 1995, p. 32.)

Another high-payoff technology is known as global broadcast services (GBS), which is analogous to commercial direct broadcast television. GBS will "enable images, large data files and other high-bandwidth information to be disseminated quickly to a number of users within a designated geographical area." Demonstration systems on UFO communications satellites will deliver 24 Mbps streams of information to fixed ground-based and mobile users within an approximately 500-nmi area using 22-inch receiver antennas. (Scott, 1996, p. 61.)

Those who champion truly ubiquitous computing envision hundreds of wireless computers throughout the home and office providing everything from "scrap computers" (functioning like scrap paper) to yard-sized displays (essentially networked chalkboards) (Weiser, 1991, pp. 98–100). Today, 10 Mbps wireless local-area networks are becoming available, and experimental systems are being developed to demonstrate even greater bandwidth capabilities (Cerf, 1991, p. 79). However, advocates recognize that the need for bandwidth to support truly ubiquitous systems will be substantial. An example from Mark Weiser of the Xerox Palo Alto Research Center (PARC) serves to illustrate (Weiser, 1993, p. 71). Imagine an office building with 300 people each possessing 100 wireless devices (entirely plausible, considering a typical work area or home has many more information sources than that). If each device demanded 256 Kbps, the aggregate bandwidth would be 7.5 Gbps. This is difficult to achieve with current wireless technologies and does not even begin to address issues associated with device mobility, power consumption and density, display size, etc. (Weiser, 1993, p. 71.)

ADVANCED PROCESSING

Processing in its broadest context runs the gamut from character recognition systems to data handling to expert decision aids. Processing capability derives to a great extent from software designed to handle tasks ranging from those that are deterministic to those that are stochastic (i.e., involve a measure of randomness). In general, it can be said that future processing capabilities will assist the virtual staff by accomplishing much of the workload it must now do faster, more accurately, and with greater "vision." Also, the computer will come to play a much more active role through designs that make it more of a collaborator with the user. Programmers will make machines appear intelligent by endowing them with certain reasoning capacities that will make them seem rather like a human assistant. (Tesler, 1991, pp. 86–87.)

Glimpses of the future can be seen in the research advances of today:

> Transcription software . . . has already appeared that enables portable machines to display printed versions of handwritten records and polished renderings of roughly drawn figures. Related software recognizes hand-drawn symbols that each represent an entire command or idea. Speech-recognition software that obeys spoken commands is also finding wider application as the technology improves. . . . Software, indeed, will change more than any other element in the computing paradigm. (Tesler, 1991, p. 90.)

Automated, intelligent optical character recognition is critical to improving work flow and reducing costs. Today, the "technology includes everything from simple bar codes to systems that can recognize handwritten upper and lower case characters, numbers and symbols." Even at only 40-percent accuracy, current OCR capabilities begin to approach humans, who at their best achieve only a 60-percent accuracy rate when reading handwritten upper and lower case letters. (Ackerman, 1994a, p. 72.) When tied to global network tracking systems, such technologies could revolutionize the means by which logistical support is provided to the warfighter.

In the field of database management, performance measured in just hundreds of transactions per second (tps) in the 1980s has advanced into the thousands today. As in other realms of computing, this progress has been accompanied by a reduction in the cost per transaction per second. Database systems capable of 10,000 tps are now on the drawing boards (Lewis, 1994, pp. 60–61). As performance levels increase, so will the prospect for timely processing and dissemination of critical target, threat reconnaissance, and BDA information (USAF Wright Laboratory, 1994).

An example of how these technologies could find their way into the ensemble of tools available to the warfighting staff is found in an effort called Footlocker, sponsored by the Air Force Rome Laboratory:

> The electronic Footlocker aids in assembly and storage of data files in a variety of areas that can meet the requirements of a deploying unit. In preparation for deployment, there may be only general or partial information for the area of operations . . . . Whatever data are available can be loaded in the system for later refinement in the theater. . . . The average soldier will be able to operate Footlocker. Once in the theater, the system will link via satellite communications for specific data to be downloaded and manipulated. (Robinson, 1994, p. 25.)

In the Air Force, Rome Lab is at the forefront of the development of many staff support technologies and has established an aggressive plan to apply them to the warfighters' needs. Rome is investing in automated decision support tools to reduce decision cycles within Air Operations Centers. It is developing "intelligent information services and software tools that will support rapid Course of Action generation; collaborative planning and execution; and enhanced visibility into the battlefield." (USAF Rome Laboratory, 1994.) Artificial intelligence techniques and other analytical tools are being applied to permit machines to read free text and to update databases automatically. Research to develop techniques to detect changes in imagery autonomously, locate targets precisely, fuse the results with other sensor data, and provide real-time dissemination for BDA is being funded. The lab even signed up to the lofty goal of reducing signals intelligence analysis "from weeks to minutes." (USAF Rome Laboratory, 1994.)

Other researchers are concentrating on providing means to help computer operators manipulate, organize, and digest the enormous amounts of information that will be generated on battlefields of the future. To do this, "software agents" may become the next "hot" technology. Such agents span a wide spectrum in form and capability, but are essentially "smart programs that can engage and help all types of computer users with their tasks." Current versions of this concept are already making an impact as data filters and information finders. (Juliussen, 1995, p. 46.)

Dr. Jude E. Franklin, vice president and chief technology officer for PRC Incorporated, describes a similar concept he calls the "information navigator." The information navigator provides "seamless, interactive, user-driven access to the mass of information within the national infrastructure." Research in the development of this technology has the practical objective of greatly enhancing the rate at which users can retrieve and assimilate information. The navigator starts by helping locate and gather the *right* information and then "visualizing" it for the user in such a way that it is quickly understood. (Robinson, 1994, pp. 25–26.)

Still another concept along the same lines as the information navigator is the Knowbot™ (a trademark of the Corporation for National Research Initiatives). Knowbots™ are programs designed to travel through a network, inspecting and understanding information, regardless of the language or form in which it is expressed. (Dertouzos, 1991, p. 67.) These programs move from machine to machine across a network, potentially with the ability to clone themselves. They achieve their objectives quickly by communicating with one another, with various servers in a network and with users and by supporting parallel computations at different sites. (Cerf, 1991, p. 74.)

However, the techniques that may pace information-processing capabilities in the future are those found in the realm of expert systems. Such technologies as fuzzy logic, AI, or neural networks may finally realize their potential with the advent of networks and massively parallel architectures. This "union" will permit the creation of "neural computers using the same principles of information processing that a biological nervous system uses." Conventional computers use inflexible programs, while neural systems are capable of self-organization and learning. "Self-organization implies structures that can configure themselves and provide correct answers, driven only by the temporal and spatial correlations that the data itself contains." By design, neural systems are built to operate autonomously in an unpredictable world. Neural networks mimic the way in which the brain is believed to work, but with the speed and accuracy of a machine. (Faggin, 1992, p. 134; Tesler, 1991, p. 92.)

But is machine intelligence really possible? From the perspective of basic microprocessor support for such a capability, it is worth noting the following argument, made by Dr. Gregory H. Canavan, senior scientific advisor at Los Alamos National Laboratory. As Dr. Canavan relates, current high-end military computers might contain chips possessing a few million transistors and performing about 300 million computations per second. Transposing this to biological terms puts today's computing machines somewhere between a worm and a bee on a geometric curve. Computers matched the worm about 20 years ago and will match the bee in another 20. Assuming such a progression rate continues, computers some 40 years from now will have a million times today's capacity, or about a trillion connections (10^{12}). In comparison, the human brain possesses about 10^{11} neurons, each of which has about 10^4 interconnections. This results in a system complexity of about 10^{15} connections. Accordingly, even in the fourth decade of the 21st century, humans might still be 10^3 (or 1,000) times more capable than the best computers of that era. Notwithstanding arguments that fast parallel-processing techniques may allow machines to close the gap, Dr. Canavan says that, for much of the intervening time,

> humans will still be the dominant element in all military machines. . . . The key to success will be routing information into and out of them as fast as possible and adding all of the virtual reality and other support systems available to assist their decisions. (Canavan, 1993, p. 50.)

In the meantime, "the basic information-processing principles that the brain uses will begin to yield to human scientific inquiry," and such knowledge will eventually "enable humans to build truly intelligent, autonomous machines." (Faggin, 1992, p. 134.)

DISPLAYS AND PRESENTATION

Regardless of how fast information is gathered, processed, and delivered, it must still be interpreted by those who deal with it. Thus, considering the complexity and enormity of the data future military staffs will have to deal with, advances in information displays and presentation formats will be required. The goal of system development in this regard will be to "provide natural human interface to complex C^3I systems through the integration of multimedia, virtual reality, high resolution displays, and spoken language." (USAF Rome Laboratory, 1994.)

In computer displays, this year will see such critical improvements as the dual-domain liquid crystal panel (Juliussen, 1995, p. 47). By the turn of the century, it is probable that a 1,000 by 800–pixel, high-contrast display will be less than 1 cm thick and weigh under 4 oz. Advances in power-supply design and in batteries will eventually enable several days of continuous use without recharging (Weiser, 1991, p. 101).

Active-matrix liquid-crystal display technology will continue its amazing progress and, according to one expert, will permit construction of wall-size, flat-panel televisions by the end of the decade. The typical classrooms of the 21st century will use displays in place of blackboards. Cathode ray tubes will also become obsolete with the maturing of amorphous silicon and polysilicon thin-film, transistor-driven, flat-panel displays. (Shur, 1993, pp. 103–104.)

While the above technologies will certainly lead to crisper images with greater resolution at lower cost, software interfaces must be used to organize and present information visually. One proposed means for doing this represents an evolution from the Windows format: the "Information Visualizer," developed by Xerox PARC. The Information Visualizer uses

> 3-D real-time animation to present information as 3-D interactive objects. Within the Information Visualizer, work is distributed throughout a collection of 3-D (and 2-D) rooms furnished with interactive objects such as walls and floating trees. . . . Rooms help manage information once it has been retrieved. By making it inexpensive to switch to a new set of windows and objects, Rooms encourages the user to keep less clutter on the screen. (Clarkson, 1991, pp. 277–282.)

Critical to this scheme is the use of animation. As PARC has learned, animation shifts workload to the user's "unconscious perceptual system, freeing the conscious mind for higher-level cognitive work." The use of natural perceptual cues "draws" the user in as an aid in assimilating and using information. Also, the Information Visualizer constantly engages the user by never forcing him to sit idle while delay prompts block input commands in favor of processing. (Clarkson, 1991, p. 281.)

When size constraints become just too small to permit adequate assimilation of information, as might be found in mobile or portable systems, helmet-mounted virtual-reality displays might provide a solution. In such devices, the image displayed to the eyes shifts as the user's head turns to give the sensation of a surrounding screen. This effectively increases the screen area available for image presentation. A natural extension of this concept would be to network this equipment by way of broadcast technologies. One advocate predicts that, by 2000, soldiers could be wearing head-up display helmets that let them receive remote sensor imagery to see over the horizon. (Lewis, 1994, p. 61.) But lest these dizzying prospects get anyone's optimism up too high, it appears that real-time holography will not be showing up in the living room at least for several more decades (Negroponte, 1991, p. 110).

Small devices might also have to dispense with the standard keyboard and instead use handwriting pads for user input. Today's personal digital assistants (PDAs) represent the current implementation of this technology (Forman and Zahorjan, 1994, pp. 45–46). Eventually, spoken-command recognition systems might one day eliminate the need for the user to be in physical contact with or even to see his computer or other network interface.

MOBILITY

A hallmark of military forces is their mobility, and wherever military forces have gone, so has their leadership in the form of staffs. While the technologies supporting the concept of virtuality may greatly reduce the size of deploying staffs, it is unlikely they will ever be eliminated completely. In the future, much of the information necessary for staff functions will come directly from the battlefield or its supporting train. Accordingly, one might predict that personnel charged with conducting operations—whether on the battlefield or in the rear—will be co-opted into staff decision cycles. Thus, without a doubt, technologies supporting staff mobility will gain in importance.

The potential for fiber-optic support of deployed forces has already been alluded to and may well become a reality via the proliferation of this technology. However, wireless technologies will continue to have a feature role in C^{4}I systems. In the mobile environment, an obvious, but non-trivial, requirement of computing systems is the ability to access information regardless of location. (Satyanarayanan, 1993, p. 81.) Unfortunately, many constraints currently affect the design of mobile systems. "[W]ireless communication is characterized by [relatively] lower bandwidths, higher error rates, and more frequent spurious disconnections." (Forman and Zahorjan, 1991, p. 39.) The number of mobile devices in a network cell (i.e., the area covered by a transceiver such as a satellite) varies over time, with the possibility of large concentrations of mobile users, which can overload network capacity. Mobile computers also face increased risks for "physical damage, unauthorized access, loss, and theft." (Forman and Zahorjan, 1994, pp. 39, 44.)

Another problem has to do with the more general concept of networking in the mobile environment. Over the past two decades, networking has been developed on the assumption that a computer's name and network address were fixed. Mobile computers introduce several problems, including dynamic changes in a mobile computer's network address, the effects of current location on configuration parameters and on answers to user queries, and growth in the communication path as the mobile user wanders away from a nearby server. (Forman and Zahorjan, 1994, p. 42.)

This is an immediate challenge with no historical precedent. It is thus difficult to predict how technology will be used to deal with the issues. Nevertheless, some general observations can be made. Designers of mobile computers will be striving for the properties characteristic of a wristwatch: small, lightweight, durable, operable under a wide variety of environmental conditions, and using minimal power, to prolong battery life. Batteries are the largest single source of weight in a portable computer, and advances in power-related technologies (from storage devices to circuit designs minimizing power demand) will be required. (Forman and Zahorjan, 1994, p. 43.)

Sophisticated dynamic bandwidth reservation and multicast algorithms will have to be developed. This is because "fiber is point-to-point, whereas satellite and ground-based radio are broadcast and multicast." (Kleinrock, 1991, p. 115.) Additionally, mobile networks must be designed to be robust in the face of remote site failures and other spurious interruptions. Sophisticated network management and processing applications must be created to sense and deal with unique circumstances while minimizing user involvement. The goal will be to offer the user the best service attainable at the current physical location and in the current network environment. For example, when limited bandwidth is available, a demand for full-motion video could be met with black-and-white rather than higher bandwidth color. (Satyanarayanan, 1993, pp. 81–82; Forman and Zahorjan, 1994, p. 41.)

The near-term will probably see expansion of the VTC capabilities first seen in the Haitian operation. Next will be a proliferation of communications devices down to the individual weapon platform or soldier. As related by the Speaker of the House, Newt Gingrich: "Early in the next century, cellular telephones will be available on the battlefield, and soldiers will be calling home from there." (Mowery, 1994, p. 61.) Perhaps more ambitiously, Air Force Lt Gen Edmonds sees the day when "even individual soldiers on the battlefield will be able to transmit and receive video imagery. Linked by communications satellites, VTC could take place anywhere on the globe." The general also spoke of a recent demonstration in which a video camera mounted on a soldier's rifle was able to send real-time surveillance imagery back to a command center and to other units. (Cooper and Holzer, 1995, p. 2.)

In parallel, advanced C^4I technologies will weave such platforms as AWACS and JSTARS and such systems as JTIDS together as the constituent elements to build an overarching view of the battlefield. Next will be the real-time fusion of this information with almost instantaneous delivery to all command staff locations. When staff locations themselves are maneuvering entities, then the dream of true mobility will have been realized.

SECURITY

Of all the challenges system designers face as they go about building the infrastructure to support the military infosphere, the greatest is the never-ending requirement for security. The vast majority of military communications involve some level of sensitivity, and future information systems will have to operate with this need in mind. Security must pervade the infosphere in light of the fact that tomorrow's heavily networked systems will permit many possible points of entry to gain unauthorized access. To paraphrase House Speaker Gingrich once again,

> the United States should be prepared for zones of creativity in opponents that it has never dreamed of, because no assurance exists that the use of cyberspace will be in any way a monopoly. (Mowery, 1994, p. 61.)

This theme of a panoply of threats to system integrity was echoed by speakers at the Armed Forces Communications Electronics Association, Europe, Rome Symposium and Exposition held in May 1994. It was pointed out that, "because total security probably is an unachievable goal, threats and risks must be faced by an adapted management approach." These experts went on to discuss securing multiplatform environments by gaining user community acceptance of multilevel security. (Schaeffer, 1994, p. 63.)

Progress in information technology is not only occurring at a rapid pace, it is occurring for the most part within the commercial sector. One must presume, then, that this technology is available to potential adversaries as much as it is to U.S. and allied forces. (Druffel et al., 1994, pp. xxv–xxvi.) A consolation may be that the commercial world also has its own security needs and thus is not uninvolved, as evidenced by interest in encryption devices, such as the "Clipper" chip. Nevertheless, recent problems with "hackers" and their successful penetration of a number of government and commercial systems may be a preview of the kinds of threats network operators will have to deal with in the future. (Bankes, 1992, p. 22.)

The implementation of multiple servers that coordinate with each other and networks that can be treated as one large disk drive is a tremendous advantage to intended users, but an invitation to unauthorized access. As "transparent data access across mainframes, minicomputers, and personal computers" becomes reality, connectivity will improve to the point that anyone with access to the system will be able to view and potentially manipulate records and tables that are located on an individual computer's file server. (Nance, 1991, pp. 268–269.) If access is gained to one part of the system, all other parts could become vulnerable to compromise.

Knowbot™-type programs (described above) would make ideal intelligence agents. They could be programmed to voyage through networks, especially those with transparent access, to gather data without restriction or to manipulate files to achieve disinformation. Knowbots™ could then "report back" to their point of origin on the success of their missions, bring back vast amounts of sensitive information, or simply inform their creators of the data available to an adversary.

To illustrate the seriousness of this threat, one need only note the reported increase in the number of intrusions by "crackers"—programmers' jargon for people who maliciously meddle with computer systems. According to a 1992 USA Research, Incorporated, report, the "number of unauthorized intrusions detected in the U.S. workplace grew from 339,000 in 1989 to 684,000 in 1991," or over 100 percent in two years (Roush, 1995, p. 37). The Computer Emergency Response Team (CERT) was formed in 1988 by the then Defense Advanced Research Projects Agency to coordinate responses to crises like the 1988 Internet "worm." CERT received reports of "some 130 incidents in 1990, 800 in 1992, 1,300 in 1993, and 2,300 in 1994." (Roush, 1995, p. 37.) In 1994, over a million secret passwords were stolen by using "sniffer" programs that were planted in computers at dozens of Internet hubs.

The DISA admitted in July 1994 "that crackers had penetrated 'major portions' of the Pentagon's unclassified networks, 'adversely affecting' the nation's military strength." An instructor who teaches defense against "data warfare" at the National Defense University adds that "people who are doing the attacks are not just crackers. They are information terrorists. Their purpose is to damage the economic and defensive capabilities of the United States." (Roush, 1995, p. 37.) According to a 1994 report on organized crime by the Center for Strategic and International Studies, "a despot armed with a computer and a small squad of expert hackers can be as dangerous and disruptive as any adversary we have faced since World War II." (Roush, 1995, pp. 37–38.)

One example demonstrates how vulnerable current Air Force computing systems can be. The Director of Computer Crime Investigations for the Office of Special Investigations at Bolling Air Force Base, James V. Christy II, solicited help from a young hacker who had plead guilty to breaking into a Pentagon computer system. He had the hacker put in a room and closely monitored all his activities so that everything was recorded. The results were astonishing. "Within 15 seconds he broke into the same computer at the Pentagon that he was convicted for, because its administrators still had not fixed its vulnerabilities." Over a three-week period, the hacker broke into over 200 Air Force systems, with not one of the victims reporting they had been compromised. (Roush, 1995, pp. 38–39.) This lack of secure defenses has ominous implications for fully networked systems.

Such vulnerabilities are not limited to on-line access and indeed extend across a broad spectrum. Radio signals can be "intercepted, altered, and rebroadcast without the knowledge of the sender or receiver." This leads to the conclusion that authentication schemes will become necessary (Lewis, 1994, p. 63). Accordingly, a crucial area for research and development in network management concerns the security of the system at all levels (Cerf, 1991, p. 81). The problem is especially acute for "answering dynamic location queries" as this function requires knowledge of the location of a mobile user. As one can imagine, in the military context this in itself could be sensitive information (Forman and Zahorjan, 1994, p. 43).

Like the broadcast medium, cable systems are also vulnerable to unauthorized access. According to a Navy intelligence briefing, entitled "Submarine Proliferation," the burgeoning use of undersea fiber optics has led to a "growing worldwide interest in eavesdropping on this emerging means of transmitting huge quantities of data and information." Russian work to create a class of special operations subs for this very purpose is cited as an example of the threat. (Holzer, 1995, p. 3.)

To address these numerous security concerns, a number of technologies are emerging. For example, Avanti Associates has created "TAG" labels, which incorporate eye-readable and bar-code serial numbers for tracking or use magnetic strips or passive RF antennas for gateway access. A concept under continuing development is that of the firewall. A firewall "places a barrier between a computer system and outside or unauthorized access." Its main purpose is "to control access to or from a protected network," and it is usually placed as a high-level gateway, such as a node to the Internet. Advantages to firewalls include "controlled access to site systems, enhanced privacy, security policy enforcement and logging of access to the network and network usage." (Lesser, 1995, pp. 16–17.)

Encryption is another method of protecting transmitted data and access to databases. The National Security Agency (NSA) has long been the U.S. expert in this arena, and NSA's "type-1 encryption devices are coming out of the 'black' and into commercial use." At the system level, Motorola's Network Encryption System permits separation of classified from unclassified areas of a network. A concept called a *kernel* implements the actual encryption-decryption algorithm process. The kernel acts as the "master" of the classified and unclassified areas, which each have their own buses and power supplies. While the kernel is able to read from and write to each side, the separate areas cannot do so on their own. (Lesser, 1995, p. 19.)

Another technology that could very well solve the challenge of secure transmissions is known as "quantum cryptography." According to physicist Richard Hughes of Los Alamos National Laboratory,

> The security of [quantum cryptography] is based on the laws of nature. . . . Unless there's something totally wrong with our understanding of quantum mechanics, then these transmissions should be utterly secure. (Davidson, 1996, p. D8.)

Quantum cryptography rests upon the principle that certain aspects of subatomic processes are "inherently unknowable." A stream of information coded using such a technology could not be cracked either by exotic algorithms or pure computing power. Furthermore, any attempts to eavesdrop would, in and of themselves, interfere with the coded transmissions, thus exposing the (unsuccessful) efforts to gain clandestine access.

Quantum cryptography is no pipe dream. IBM researchers demonstrated it in the lab in the late 1980s using laser signals transmitted through optical fibers. The current limitation is transmission distance, which is only on the order of yards, but efforts are under way to extend the capability into the more useful range of several miles. Ultimately, the goal will be to make the technology work in free space and enable ground-to-satellite applications. As Richard Hughes sees it, "We'll know in a year or two if it works." (Davidson, 1996, p. D8.)

When considering the variety of ways information can be compromised, it may seem easy to paint a bleak picture of the dire prospects of providing for adequate security. The technologies listed above may seem modest, but they are proving effective, and they represent only a subset of what has been discussed in the open literature (never mind that available in the classified world). While the requirements for securing our communications and databases remain formidable, history has demonstrated that such challenges can be faced and met. There is no reason to doubt that the future will be any different.

COMMERCIAL NEXUS

As mentioned earlier, the commercial world is the primary engine driving the technological advancements in the infosphere. Andrew F. Krepinevich, Jr., Director of the Defense Budget Project in Washington, D.C., and adjunct professor of strategic studies at the Paul A. Nitze School of Advanced International Studies, Johns Hopkins University, has observed that the military-technological revolution in information is "highly diffused, occurring as much, if not more, in the commercial sector as in the defense sector, and throughout the advanced industrial world." (Krepinevich, 1994.) His words are buttressed by Admiral Owens' observation that the amount of research and development (R&D) the DoD sponsors as a percentage of that conducted throughout the country has greatly declined. From historical levels of 20 percent, it has dropped to the 3 to 4 percent range.[2] Admiral Owens also points out that the center of technological acceleration in each of three core areas—digitization, computer processing, and global positioning—lies primarily in the nondefense sector. This has enormous impact on how the military is to formulate its R&D and procurement practices. As the Air Force Scientific Advisory Board has stated:

> From a technology perspective, effort is needed to balance commercial off the shelf (COTS) and standards with innovation in C[4]I/information systems research to stay ahead of adversaries. The Air Force needs a paradigm shift to focus on technology transfer

---

[2]Admiral Owens, as quoted in "Exploit Civilian Technologies Says No. 2 US Military Leader," 1995, p. 19.

from the commercial sector to the military. (Druffel et al., 1994, p. xxvi.)

For the military to achieve this goal, it must accentuate such historical strengths as its "vast experience and capabilities in terms of systems integration." (Morrocco, 1995, p. 23.) Figure A.4 indicates how such strengths might be brought to bear to enhance satellite communications capabilities. This figure shows, that in the future, commercial systems might be hybridized with those of the military to increase relative individual bandwidth availability and network node connectivity. (Sorensen, 1995.)

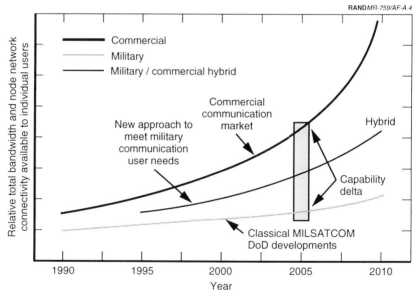

SOURCE: A. N. Sorensen, course on "An Introduction to Space Communication Architecture," El Segundo, Calif.: Aerospace Corporation, 1995. Used by permission.

Figure A.4—Potential to Leverage Commercial Satellite Communications Capability for Military Use

In the realm of security, the government is not the only organization concerned with maintaining the integrity of its databases, communications, and transmissions. Commercial institutions also require secure communications and information storage. One example is the sanctity of financial transactions. Another is the sensitivity of research and development investments to get products to market as expeditiously as possible and to minimize industrial espionage. Customer personal information is also sacrosanct. Companies recognize that those who do not protect this type of information will rapidly lose their customers and any product development advantages they possess to their competitors. Steps are being taken to minimize these security breaches right now and probably faster than if the government was the only customer.

Obviously, this shift in how the military acquires and employs its future weapon systems means military organizations are going to have to change their way of doing business. Old paradigms must give way to new. Military forces must learn to leverage commercial technologies (and managerial approaches as well) to the maximum extent possible to stay ahead of potential adversaries who are free to do the same. Acquisition strategies must become more attuned to the possibilities inherent throughout the military architecture so as to maximize return on investment in a shrinking budgetary environment.

BIBLIOGRAPHY

Ackerman, Robert K., et al., "Imaging Brightness as Key Enabling Technology," *Signal*, Vol. 48, No. 12, August 1994, pp. 71–72.

Ackerman, Robert K., "Military, Commercial Boundaries Continue to Dissolve in Silicon," *Signal*, Vol. 48, No. 12, August 1994, pp. 65–68.

Ackerman, Robert K., "Technology Solutions Beckon, Daunt Government, Industry," *Signal*, Vol. 48, No. 12, August 1994, pp. 68–70.

Armed Forces Staff College, *The Joint Staff Officer's Guide 1993*, Washington, D.C.: U.S. Government Printing Office, AFSC PUB 1, 1993.

Arquilla, John, and David Ronfeldt, "Cyberwar Is Coming," Santa Monica, Calif.: RAND, P-7791, 1992.

Bankes, Steve, et al., "Seizing the Moment: Harnessing the Information Technologies," *The Information Society*, Vol. 8, No. 1, January–March, 1992, pp. 1–59.

Barrett, Archie D., *Reappraising Defense Organization*, Washington, D.C.: National Defense University, 1983.

Barron, Janet J., "Cubelets and Chiplets," *BYTE*, Vol. 17, No. 2, February 1992, pp. 144–145.

Boar, Bernard H., *The Art of Strategic Planning for Information Technology*, New York: John Wiley & Sons, Inc., 1993.

Brody, Herb, "Internet@crossroads.$$$," *Technology Review,* May/June 1995, Vol. 98, No. 4, pp. 24–31.

Builder, Carl H., *The Masks of War: American Military Styles in Strategies and Analysis,* A RAND Corporation Research Study, Baltimore, Md.: The John Hopkins University Press, 1989.

Burger, Robert M., and William C. Holton, "Reshaping the Microchip," *BYTE,* Vol. 17, No. 2, February 1992, pp. 137–147.

Campen, Alan D. (Col USAF, Ret.), "Intelligence Leads Renaissance in Military Thinking," *Signal,* Vol. 48, No. 12, August 1994, pp. 17–18.

Canavan, Gregory H., "Changing Times Implode Defense Science Dynamics," *Signal,* Vol. 48, No. 1, September 1993, pp. 49–50.

Cerf, Vinton G., "Networks," *Scientific American,* Vol. 265, No. 3, September 1991, pp. 72–81.

Challener, Richard D., *The French Theory of the Nation in Arms 1866–1939,* New York: Columbia University Press, 1955.

Clarkson, Mark A., "An Easier Interface," *BYTE,* February 1991, pp. 277–282.

Clausewitz, Carl Von, *On War,* Princeton, N.J.: Princeton University Press, 1984.

Cole, Alice C., Alfred Goldberg, Samuel A. Tucker, and Rudolph A. Winnacker, *The Department of Defense Documents on Establishment and Organization 1944–1978,* Washington, D.C.: Office of the Secretary of Defense Historical Office, 1978.

Combined Arms and Services Staff School, *Historical Development of Staffs,* Ft. Leavenworth, Kan., E102-1, 1990.

Cooper, Pat, and Robert Holzer, "Video Teleconferencing to Join the Battlefield," *Defense News,* Vol. 10, No. 19, May 15–21, 1995, p. 2.

Cooper, Pat, and Frank Olivieri, "Hacker Exposes U.S. Vulnerability," *Defense News,* Vol. 10, No. 40, October 9–15, 1995, pp. 1, 37.

Craig, Gordon A., *The Politics of the Prussian Army 1640–1945*, New York: Oxford University Press, 1955.

Davidson, Keay, "Quantum Leap Sought for Keeping Codes Secure," *The Washington Times*, April 28, 1996, p. D8.

Delderfield, R.F., *Napoleon's Marshals*, Philadelphia: Chilton Books, 1962.

Dellecave, Tom, Jr., "Wireless Off the Ground," *Information Week*, May 29, 1995, pp. 33–34.

Denning, Peter J., "Worldnet," *American Scientist*, Vol. 77, September–October, 1989, pp. 432–434.

Dertouzos, Michael L., "Communications, Computers and Networks," *Scientific American*, Vol. 265, No. 3, September 1991, pp. 62–69.

Doherty, Dan (ed.), "U.S. Military Eyes Private Sector to Obtain Key Satellite Services," Military Newswire Service, May 16, 1995.

Druffel, L. E., et al., *Information Architectures that Enhance Operational Capability in Peacetime and Wartime*, U.S. Air Force Scientific Advisory Board, SAB 94-002, February, 1994.

"Exploit Civilian Technologies Says No. 2 US Military Leader," *The Christian Science Monitor*, Vol. 87, No. 116, May 11, 1995b, p. 19.

Faggin, Federico, "Massive Parallelism: The Name of the Game," *BYTE*, Vol. 17, No. 2, February 1992, p. 134.

Forman, G. H., and J. Zahorjan, "The Challenges of Mobile Computing," *Computer*, Vol. 27, No. 4, April 1994, pp. 38–47.

Gallaher, John G., "The Prussian Regiment of the Napoleonic Army," *The Journal of Military History*, Vol. 55, The Society for Military History, July 1991.

GAO Report, *Defense Reorganization: Progress and Concerns at JCS and Combatant Commands*, GAO/NSIAD-89-83, March 1989.

Garner, Jay M. (LGEN, USA), "Space—'The Future of Force XXI,'" *Army*, Vol. 44, No. 11, November 1994, pp. 20–24.

Geppert, Linda, "Solid State," *IEEE Spectrum*, Vol. 23, No. 1, January 1995, pp. 35–39.

Goerlitz, Walter, *History of the German General Staff*, tr. Brian Battershaw, New York: Frederick A. Praeger, 1962.

Goldman, Steven L., Roger N. Nagel, and Kenneth Preiss, *Agile Competition and Virtual Organizations*, New York: Van Nostrand Reinhold, 1995.

Grier, Peter, "Information Warfare," *Air Force Magazine*, Vol. 78, No. 3, March 1995, pp. 34–37.

Hammer, Michael and James Champy, *Reengineering the Corporation: A Manifesto for Business Revolution*, New York: Harper Business, 1993.

Hittle, J. D., *The Military Staff: Its History and Development*, Harrisburg, Penn.: The Military Service Publishing Company, 1949.

Holzer, Robert, "Russians Invest in Special Ops Subs," *Defense News*, Vol. 10, No. 19, May 15–21, 1995, p. 3.

Irvine, Dallas D., "The Origin of Capital Staffs," *The Journal of Modern History*, Vol. X, Society for Military History, June 1938.

Jain, P. C., "Architectural Trends in Military Satellite Communications Systems," *Proceedings of the IEEE*, Vol. 78, No. 7, July 1990, pp. 1177–1189.

Juliussen, Egil, "Small Computers," *IEEE Spectrum*, Vol. 32, No. 1, January 1995, pp. 44–47.

Kinder, Hermann, and Werner Hilgemann, *The Anchor Atlas of World History*, Vol. I, tr. Ernest A. Menze, New York: Anchor Books, 1974.

Kleinrock, Leonard, "ISDN—The Path to Broadband Networks," *Proceedings of the IEEE*, Vol. 79, No. 2, February 1991, pp. 112–117.

Kobb, Bennett Z., "Telecommunications," *IEEE Spectrum*, Vol. 32, No. 1, January 1995, pp. 30–34.

Korb, Laurence J., *The Joint Chiefs of Staff: The First 25 Years*, Bloomington, Ind.: Indiana University Press, 1976.

Krepinevich, Jr., Andrew F., "Keeping Pace with the Military Technological Revolution," *Issues in Science and Technology*, Summer 1994.

Landler, Mark, "It's Not Only Rock 'n' Roll," *Business Week*, October 10, 1994, pp. 83–84.

Lesser, Roger, "So You Think Your Information is Secure—Is It?" *Defense Electronics*, Vol. 27, No. 5, May 1995, pp. 16–19.

Lewis, T. G., "Where Is Computing Headed?" *Computer*, Vol. 27, No. 8, August 1994, pp. 59–63.

Lowe, S. J., "Data Communications," *IEEE Spectrum*, Vol. 32, No. 1, January 1995, pp. 26–29.

Luttwak, Edward N., *The Pentagon and the Art of War*, New York: Simon and Schuster, Inc., 1985.

Lynch, David J., "The Air Force Takes Stock," *Air Force Magazine*, Vol. 78, No. 2, February 1995, pp. 24–31.

Macedonia, Michael R. (MAJ, USA), "Information Technology in Desert Storm," *Military Review*, October 1992.

Mann III, Edward C. (Col, USAF), *Thunder and Lightning*, Maxwell Air Force Base, Ala.: Air University Press, April 1995.

Matthews, William, "How to Tie It All Together," *Air Force Times*, No. 41, May 15, 1995, p. 24.

Meadows, Sandra I., "Satellite Links Bestow Commanders with Split-Second Combat Spectacle," *National Defense*, March 1995, pp. 24–26.

"Moore's Law Meets MIPS," *BYTE*, Vol. 17, No. 2, February 1992, p. 125.

Morrocco, John D., "U.S. Military Eyes Revolutionary Change," *Aviation Week & Space Technology*, Vol. 142, No. 18, May 1, 1995, pp. 22–23.

Mowery, Beverly P., "Technology Opens New Strategies for Future Battlefield Operations," *Signal*, Vol. 48, No. 12, August 1994, p. 61.

Murray, Matt, "Thanks, Goodbye," *The Wall Street Journal*, Thursday, May 4, 1995, p. A1.

Nance, Barry, "The Future of Network Operating Systems," *BYTE*, February 1991, pp. 268–272.

Negroponte, Nicholas, "Products and Services for Computer Networks," *Scientific American*, Vol. 265, No. 3, September 1991, pp. 106–113.

Newell, Lt Col Clayton R., "The Technological Future of War," *Military Review*, Vol. LXIX, No. 10, October 1989, pp. 22–28.

Novack, Marcus, "Liquid Architectures in Cyberspace," in *Cyberspace: First Steps*, M.I.T. Press, Cambridge, Mass., 1991.

Owens, William A. (ADM, USN), "The Emerging System of Systems," *U.S. Naval Institute Proceedings*, Vol. 121/5/1107, May 1995, pp. 35–39.

Palmer, R.R., and Joel Colton, *A History of the Modern World*, Alfred A. Knopf, Inc., 1950.

Paret, Peter (ed.), *Makers of Modern Strategy from Machiavelli to the Nuclear Age*, Princeton, N.J.: Princeton University Press, 1986.

Powell, Colin L., "Information-Age Warriors," *BYTE*, Vol. 17, No. 7, July 1992, p. 370.

Reynolds, Richard T. (Col, USAF), *Heart of the Storm*, Maxwell Air Force Base, Ala.: Air University Press, January 1995.

Robinson, Jr., Clarence A., "Industry's Expanding Technology Lures Intelligence Agency Users," *Signal*, Vol. 48, No. 12, August 1994, pp. 25–27.

Roos, John G., "InfoTech InfoPower," *Armed Forces Journal International*, June 1994, pp. 31–36.

Ross, Steven T., *European Diplomatic History 1789–1815 France Against Europe*, Malabar, Fla.: Robert E. Krieger Publishing Co., 1969.

Roush, Wade, "Hackers: Taking a Byte Out of Computer Crime," *Technology Review*, April 1995, Vol. 98, No. 3, pp. 32–40.

Satyanarayanan, M., "Mobile Computing," *Computer*, Vol. 26, No. 9, September 1993, pp. 81–82.

Schaeffer, Beverly T., "Data Security Looms Crucial in Changing World Climate," *Signal*, Vol. 48, No. 12, August 1994, p. 63.

Scott, William B., "Global Broadcast Potential Explored," *Aviation Week & Space Technology*, Vol. 144, No. 6, February 5, 1996, pp. 61–62.

Sherman, Stratford, "How to Bolster the Bottom Line," *Fortune 1994 Information Technology Guide*, Autumn 1993, pp. 24–28.

Showalter, Dennis E., "Caste, Skill, and Training: The Evolution of Cohesion in European Armies from the Middle Ages to the Sixteenth Century," *The Journal of Military History*, Vol. 57, The Society for Military History, July 1993.

Shur, Michael, "Future Impact of Solid-State Technology on Computers," *Computer*, Vol. 26, No. 4, April 1993.

Sorensen, A. N., course on "An Introduction to Space Communication Architecture," El Segundo, Calif.: Aerospace Corporation, 1995.

Sproull, Lee, and Sara Kiesler, "Computers, Networks and Work," *Scientific American*, Vol. 265, No. 3, September 1991, pp. 116–123.

Tesler, Lawrence G., "Networked Computing in the 1990s," *Scientific American*, Vol. 265, No. 3, September 1991, pp. 86–93.

Van Creveld, Martin, *Command in War*, Cambridge, Mass.: Harvard University Press, 1985.

USAF Phillips Laboratory, *FY95 Space and Missiles Technology Area Plan, PL/XP*, Kirtland AFB, N.M., 1994.

USAF Rome Laboratory, *FY95 Command Control Communications & Intelligence C³I Technology Area Plan, RL/XPX*, Griffiss AFB, N.Y., 1994.

USAF Wright Laboratory, *FY95 Avionics Technology Area Plan, WL/AAOR*, Wright-Patterson AFB, Ohio, 1994.

Weigley, Russell F., "The American Military and the Principle of Civilian Control from McClellan to Powell," *The Journal of Military History*, Special Issue 57, The Society for Military History, October 1993.

Weiser, Mark, "The Computer for the 21st Century," *Scientific American*, Vol. 265, No. 3, September 1991, pp. 94–104.

Weiser, Mark, "Ubiquitous Computing," *Computer*, Vol. 26, No. 10, October 1993, pp. 71–72.

Widnall, Dr. Sheila E., and Ronald R. Fogleman (Gen, USAF), "1995 Joint Posture Hearing Statement to the Senate Armed Services Committee," Washington, D.C., April 28, 1995.

Wind, Shalom, and Theoren Smith, "Computers Take a Quantum Leap," *BYTE*, Vol. 17, No. 2, February 1992, pp. 140–141.

Winnefeld, James A., Preston Niblack, and Dana J. Johnson, *A League of Airmen: U.S. Air Power in the Gulf War*, Santa Monica, Calif.: RAND, 1994.